Harcourt Math

GEORGIA EDITION

Challenge Workbook

TEACHER EDITION
Grade 3

Harcourt

Visit The Learning Site!
www.harcourtschool.com

Copyright © by Harcourt, Inc.

All rights reserved. No part of this publication may be reproduced or transmitted in any form or by any means, electronic or mechanical, including photocopy, recording, or any information storage and retrieval system, without permission in writing from the publisher.

Requests for permission to make copies of any part of the work should be addressed to School Permissions and Copyrights, Harcourt, Inc., 6277 Sea Harbor Drive, Orlando, Florida 32887-6777. Fax: 407-345-2418.

HARCOURT and the Harcourt Logo are trademarks of Harcourt, Inc., registered in the United States of America and/or other jurisdictions.

Printed in the United States of America

ISBN 13: 978-0-15-349559-5
ISBN 10: 0-15-349559-6

If you have received these materials as examination copies free of charge, Harcourt School Publishers retains title to the materials and they may not be resold. Resale of examination copies is strictly prohibited and is illegal.

Possession of this publication in print format does not entitle users to convert this publication, or any portion of it, into electronic format.

1 2 3 4 5 6 7 8 9 10 054 15 14 13 12 11 10 09 08 07 06

CONTENTS

▶ Unit 1: UNDERSTAND NUMBERS

▶ **Chapter 1: Number Sense and Place Value**
- 1.1 Lucky Number 1
- 1.2 Colorful Balloons 2
- 1.3 Arrays of Stars 3
- 1.4 Problem Solving Strategy: Use Logical Reasoning 4
- 1.5 Pattern Plans 5
- 1.6 Matching Numbers 6

▶ **Chapter 2: Compare, Order, and Estimate**
- 2.1 Missing Numbers 7
- 2.2 Model Numbers 8
- 2.3 Scrambled Numbers! 9
- 2.4 Table Talk 10
- 2.5 Quick Sale 11
- 2.6 Round and Round We Go! 12

▶ Unit 2: ADDITION, SUBTRACTION, MONEY, AND TIME

▶ **Chapter 3: Addition**
- 3.1 Colorful Matches 13
- 3.2 Missing Addend Riddle 14
- 3.3 Nature's Numbers 15
- 3.4 Addition Bubbles 16
- 3.5 Palindromes 17
- 3.6 Get In Shape 18
- 3.7 Add Greater Numbers 19
- 3.8 Write Expressions and Number Sentences 20

▶ **Chapter 4: Subtraction**
- 4.1 Estimate Differences 21
- 4.2 Subtraction Bubbles 22
- 4.3 Subtract 3- and 4-Digit Numbers .. 23
- 4.4 Missing Digits 24
- 4.5 Planning a Party 25

▶ **Chapter 5: Use Money**
- 5.1 Colorful Sets 26
- 5.2 Paying Cash 27
- 5.3 Shopping at the Pet Store 28
- 5.4 Make Change 29
- 5.5 Money Matters 30

▶ **Chapter 6: Understand Time**
- 6.1 Find the Time 31
- 6.2 Time for a Riddle 32
- 6.3 Time Flies 33
- 6.4 Problem Solving Skill: Make a Schedule 34

▶ Unit 3: MULTIPLICATION CONCEPTS AND FACTS

▶ **Chapter 7: Understand Multiplication**
- 7.1 Multiply in the Sky 35
- 7.2 Pattern Plot 36
- 7.3 Dance on Arrays 37
- 7.4 Puzzling Products 38
- 7.5 What's the Question? 39

Chapter 8: Multiplication Facts Through 5
- 8.1 Fit Feasting Facts 40
- 8.2 Pondering Products 41
- 8.3 Problem Solving Strategy: Look for a Pattern 42
- 8.4 The Factor Game 43
- 8.5 Find Those Factors 44

Chapter 9: Multiplication Facts and Strategies
- 9.1 The Array Game 45
- 9.2 Number Patterns 46
- 9.3 Square Time 47
- 9.4 Finding Factor Pairs 48
- 9.5 Row after Row 49

Chapter 10: Multiplication Facts and Patterns
- 10.1 Combination Challenge 50
- 10.2 What's the Rule? 51
- 10.3 Missing Factors 52
- 10.4 Property Match Game 53
- 10.5 Special Delivery 54

Unit 4: DIVISION CONCEPTS AND FACTS

Chapter 11: Understand Division
- 11.1 Paintbrush Division 55
- 11.2 Animal Division 56
- 11.3 Missing Numbers 57
- 11.4 Fact Family Patterns 58
- 11.5 Picture Maker 59

Chapter 12: Division Facts Through 5
- 12.1 Favorite Numbers 60
- 12.2 The Same and Different 61
- 12.3 Writing Number Sentences 62
- 12.4 Write a Problem 63
- 12.5 Solving Problems at the Aquarium 64

Chapter 13: Division Facts Through 10
- 13.1 Fun with Facts 65
- 13.2 Dot-to-Dot Division 66
- 13.3 The Quotient Game 67
- 13.4 What's the Cost? 68
- 13.5 What Number Am I? 69

Unit 5: DATA AND MEASUREMENT

Chapter 14: Collect and Graph Data
- 14.1 Sara's System 70
- 14.2 You Decide 71
- 14.3 What's the Question 72
- 14.4 What is Left? 73
- 14.5 Pay Up 74

Chapter 15: Customary and Metric Length
- 15.1 No Rulers Allowed! 75
- 15.2 Choose the Best Unit 76
- 15.3 Graphing Length 77
- 15.4 Centimeter Estimation Game 78
- 15.5 What's the Order? 79
- 15.6 Take a Hike 80

Unit 6: GEOMETRY AND PATTERNS

Chapter 16: Plane Figures
16.1 What's the Angle? 81
16.2 Mapmaker, Mapmaker,
       Make Me a Map! 82
16.3 Polygon Puzzle 83
16.4 Triangle Tally 84
16.5 Quadrilateral Puzzles 85
16.6 What's Next? 86
16.7 Missing Labels 87

Chapter 17: Solid Figures
17.1 Solid Match 88
17.2 The Missing Half 89
17.3 Folding Solid Figures 90
17.4 Different View 91

Chapter 18: Perimeter and Area
18.1 Find the Perimeter 92
18.2 What's Missing? 93
18.3 Areas in Town 94
18.4 Find Area 95
18.5 Painting Project 96

Chapter 19: Algebra: Patterns
19.1 Continue the Pattern 97
19.2 Missing Tiles 98
19.3 Traveling Birds 99
19.4 Missing Numbers 100

Unit 7: FRACTIONS AND DECIMALS

Chapter 20: Fractions
20.1 Fetching Fractions 101
20.2 Color the Apples 102
20.3 Fraction Squares 103
20.4 Fraction Puzzle 104
20.5 Find the Difference 105
20.6 Solve the Riddle 106
20.7 Musical Math 107

Chapter 21: Decimals
21.1 Riddlegram! 108
21.2 Add It Up 109
21.3 Sum Match 110
21.4 Decimal Differences 111
21.5 Digit and Decimals 112

Unit 8: MULTIPLY AND DIVIDE BY 1-DIGIT NUMBERS

Chapter 22: Multiply by 1-Digit Numbers
22.1 Multiplication Patterns 113
22.2 In the Shade 114
22.3 Where's the Match? 115
22.4 Pick a Pair 116
22.5 Rules and More Rules 117
22.6 Planning a Picnic 118
22.7 Picture This! 119
22.8 Can You Find Me? 120

Chapter 23: Divide by 1-Digit Numbers

23.1 Sharing Marbles **121**
23.2 Star Quotients **122**
23.3 Arranging Digits **123**
23.4 Colorful Quotients **124**
23.5 The Division Race **125**
23.6 Divide and Check **126**
23.7 Something's Fishy! **127**

Lucky Number

Use the clues to shade the numbers in the chart.
Shade all the numbers you land on when skip-counting.
Use the chart to answer the question.

1. Skip-count by threes.
   Stop at number 36.

2. Skip-count by fives.

3. Start with the number 7.
   Skip-count by fours.

4. Start with the number 1.
   Skip-count by sevens.

5. Skip-count by twos.

6. Skip-count by eights starting with 41.

7. Start with the number 17.
   Skip-count by tens.

Which number is left unshaded? _____13_____

1	2	3	4	5	6	7	8	9	10
11	12	13	14	15	16	17	18	19	20
21	22	23	24	25	26	27	28	29	30
31	32	33	34	35	36	37	38	39	40
41	42	43	44	45	46	47	48	49	50

Name _____

LESSON 1.2

Colorful Balloons

Read the clues. Color the balloons.

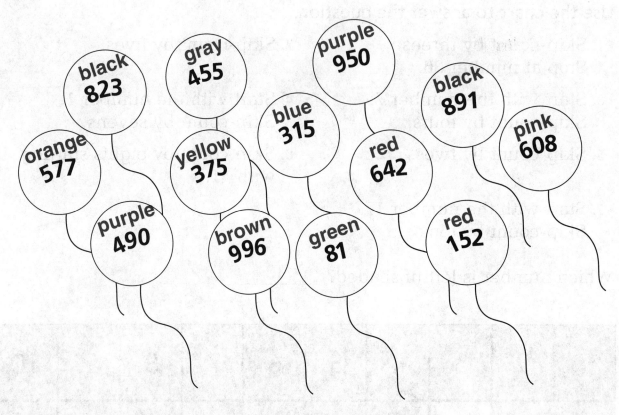

1. Any balloon with a two in the ones place is red.
2. If the balloon has a one in the tens place, color it blue.
3. A balloon whose number equals 175 + 200 is yellow.
4. The balloon with the least number is green.
5. Color any balloon with a zero in the ones place purple.
6. The balloon with the greatest number is brown.
7. Any balloon whose number is 576 + 1 is orange.
8. Any balloon with an eight in the hundreds place is black.
9. If the balloon has a zero in the tens place, color it pink.
10. If the number is 555 − 100, color the balloon gray.

CW2 Challenge

Name _____

LESSON 1.3

Arrays of Stars

Each sheet of stickers contains 1,000 stars. The stars are arranged in 50 rows with 20 stickers in each row. Use this information to answer each question.

1. A package of stickers contains 5,000 stars. How many sheets of stars are there?

   _____**5 sheets**_____

2. Each teacher at Lakota Elementary received 9 sheets of stars. How many stars did each receive?

   _____**9,000 stars**_____

3. Mr. Lee used 3,894 stickers during the first grading period. How many sheets was this?

   _____**4 sheets**_____

4. If each sheet of stickers costs 50¢, how much would 12,000 stars cost?

   _____**$6**_____

5. Mrs. Winslow used 2 sheets of stickers during the first grading period, 3 sheets during the second grading period and half of a sheet during the first week of the third grading period. How many stars did she use?

   _____**5,500 stars**_____

6. Carrie used 8 rows of stars to decorate thank-you notes for the parents that helped on the last field trip. How many stars did she use?

   _____**160 stars**_____

7. Martin made a design with the stars. He used one sheet of stars in the first row of his design, two sheets in the next row, three sheets in the third row, and so on. If his design had 6 rows, how many stars did he use?

   _____**21,000 stars**_____

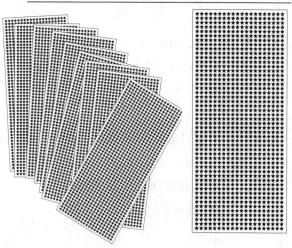

Challenge   CW3

LESSON 1.4

Name _____

Problem Solving Strategy
Use Logical Reasoning

Read the clues. Color the squares on the hundred chart.

1	2	3	4	5	6	7	8	9	10
11	12	13	14	15	16	17	18	19	20
21	22	23	24	25	26	27	28	29	30
31	32	33	34	35	36	37	38	39	40
41	42	43	44	45	46	47	48	49	50
51	52	53	54	55	56	57	58	59	60
61	62	63	64	65	66	67	68	69	70
71	72	73	74	75	76	77	78	79	80
81	82	83	84	85	86	87	88	89	90
91	92	93	94	95	96	97	98	99	100

1. A 2-digit number the sum of whose digits is 1. **10**

2. The last number on the hundred chart. **100**

3. The number that is 99 less than the greatest number on the hundred chart. **1**

4. Both digits are odd. The tens digit minus the ones digit is 4. Their sum is 10. **73**

5. The digits are equal. The sum of the digits is 10. **55**

6. The number in the sixth column, fifth row. **46**

7. The number whose digits have a sum of 15 and a difference of 1. This number is not 8 less than 95. **78**

8. If you subtract 6 from the ones digit, you get the tens digit. Skip-count by 2 four times to get the ones digit. **28**

9. The number that is 5 less than the answer to Problem 8. **23**

10. The number whose ones digit is 8 less than its tens digit. The sum of the digits is a 2-digit number. **91**

11. The number whose digits have a sum the same as the digits in 36. This number is less than 50 and greater than 36. **45**

12. The number that is 1 less than the number whose digits have a sum of 12 and whose tens digit is greater than 4 but less than 6. **56**

CW4 Challenge

Name _____

LESSON 1.5

Pattern Plans

Predict the next three numbers in each pattern. Describe the pattern.

1. 957, 947, 937, 927, __917__, __907__, __897__
   Pattern: __numbers decrease by 10__

2. 132, 137, 142, 147, __152__, __157__, __162__
   Pattern: __numbers increase by 5__

3. 824, 821, 818, 815, __812__, __809__, __806__
   Pattern: __numbers decrease by 3__

4. 356, 360, 364, 368, __372__, __376__, __380__
   Pattern: __numbers increase by 4__

5. 640, 638, 636, 634, __632__, __630__, __628__
   Pattern: __numbers decrease by 2__

Make your own patterns by increasing or decreasing 3-digit numbers. Describe each pattern.
Patterns will vary. Check students' work.

6. ____, ____, ____, ____, ____, ____, ____
   Pattern: _____

7. ____, ____, ____, ____, ____, ____, ____
   Pattern: _____

Challenge CW5

Name _____

> **LESSON 1.6**

Matching Numbers

Match the correct standard form to the word form of the number.

1. eighty-three thousand, nine hundred seventy-six

2. four thousand, three hundred four

3. sixty-two thousand, five hundred ten

4. five thousand, eight hundred twenty-six

5. forty-seven thousand, four hundred eight

6. five thousand, two hundred forty-seven

7. fourteen thousand, six hundred nineteen

8. sixty-six thousand, one hundred fifty-two

9. twenty-eight thousand, nine hundred eighty-three

10. forty-two thousand, seven hundred six

A. 14,619
B. 4,304
C. 5,826
D. 5,247
E. 83,976
F. 66,152
G. 42,706
H. 62,510
I. 47,408
J. 28,983

CW6 Challenge

Name _____

> LESSON 2.1

Missing Numbers

Detective Casey needs to find some missing numbers. Can you help her? When you find them, circle the numbers so she knows where to look. The numbers can be found going up, across, down, backward, and diagonally.

1	2	8	6	7	4	3	0	5
5	0	1	9	4	2	1	8	9
2	2	7	8	0	1	3	6	4
4	8	7	5	2	6	0	4	3
7	5	1	4	6	2	1	0	8
4	6	9	8	2	0	3	4	7
9	8	5	2	4	3	0	1	6
2	9	3	7	4	6	5	0	9
1	7	2	8	3	9	6	4	5

1. 17,283
2. 50,194
3. 3,756
4. 27,423
5. 44,104
6. 4,301
7. 9,820
8. 5,469
9. 94,387
10. 9,132
11. 6,150
12. 7,402
13. 31,301
14. 28,674
15. 5,820

Detective Casey thanks you for all your hard work in helping her find the missing numbers.

Challenge

Name _____

LESSON 2.2

Model Numbers

Look at the models below. Under each, write the number that the model represents. When you finish, follow the directions at the bottom of the page.

| ○ = hundreds | ☆ = tens | ☐ = ones |

1. ☆ ☆ ☆ ☆
   ☐ ☐ ☐ ☐
   ____44____

2. ☆ ☆ ☆
   ☐ ☐
   ____32____

3. ☆ ☆ ☆
   ☐
   ____31____

4. ☆ ☆ ☆ ☆ ☆
   ☐ ☐ ☐
   ____53____

5. ☆ ☆ ☆ ☆
   ☐ ☐ ☐ ☐ ☐ ☐
   ____46____

6. ☆ ☆
   ☐ ☐
   __|22|__

7. ○ ○ ○
   ☆ ☆
   ☐ ☐ ☐
   ____323____

8. ○ ○ ○
   ☆ ☆ ☆
   ____330____

9. ○ ○ ○
   ☆ ☆ ☆
   ☐ ☐
   __(332)__

10. In Exercises 1–9, circle the answer that is the greatest number.

11. In Exercises 1–9, what is the number that is closest in size to the greatest number?
    ____330____

12. In Exercises 1–9, put a square around the answer that is the least number.

13. In Exercises 1–9, what is the number that is closest in size to the least number?
    ____31____

14. Choose your own number. Write what that number is, and model it by using circles, stars, and squares.

    Check students' answers.

CW8 Challenge

Name _____

▶ LESSON 2.3

Scrambled Numbers!

Pick numbers from each circle to make three different 4-digit numbers. Write your numbers on the lines to the left of the circle. Then, on the lines to the right of the circle, put your numbers in order from least to greatest. **Students' answers will vary for all problems.**

1. _____ (1 6 3 / 7 8 4) _____
   _____   _____
   _____   _____

2. _____ (9 0 4 / 1 2 4 / 5) _____
   _____   _____
   _____   _____

3. _____ (6 7 2 / 3 4 1) _____
   _____   _____
   _____   _____

4. _____ (1 6 3 / 7 9 0) _____
   _____   _____
   _____   _____

5. _____ (2 4 1 / 8 6 3) _____
   _____   _____
   _____   _____

6. _____ (3 6 9 / 2 1 4) _____
   _____   _____
   _____   _____

7. _____ (9 5 0 / 2 1 3) _____
   _____   _____
   _____   _____

8. _____ (3 1 5 / 6 8 7) _____
   _____   _____
   _____   _____

9. What is the greatest number you came up with?

   Answers will vary.

10. What is the least number you came up with?

    Answers will vary.

Challenge   CW9

LESSON 2.4

Name _____

Table Talk

Carl's Pen Store expects a busy year. The table shows how many pens Carl has in his store.

Carl's Pen Store	
Blue	29,492
Red	13,650
Green	9,834
Black	6,500
Purple	2,173

For 1–8, use the table to solve.

1. Michelle buys 1,000 red pens. How many red pens does Carl have left?

   12,650 pens

2. If Carl sells all his green, black, and purple pens, how many pens will he sell in all?

   18,507 pens

3. John needs 2,000 purple pens. Are there enough at Carl's Pen Store?

   yes

4. Carl has more than 10,000 pens of which colors?

   blue and red

5. Tony orders 2,000 of the purple pens. Carl mails them in groups of 1,000. How many groups does Carl mail to Tony?

   2 groups

6. Amy needs to buy 7,000 black pens. She goes to Carl's store and buys all his black pens. How many pens does she still need to buy?

   500 pens

7. Eric needs 2,000 green pens and 1,000 red pens. How many pens does he need in all?

   3,000 pens

8. Carl mails all his blue pens to California. He can only fit 10,000 pens in a box. How many boxes will he need?

   3 boxes

CW10 Challenge

Name _____

LESSON 2.5

Quick Sale

George's General Merchandise is having a sale. Each sale table will hold items that have been estimated to the nearest ten or hundred dollars for a quick sale. Under each item, write the table it should go on.

1.  $189

   Table F

2. $25

   Table B

3.  $175

   Table F

4.  $23

   Table A

5.  $187

   Table F

6.  $259

   Table G

7.  $37

   Table C

8.  $157

   Table F

9.  $313

   Table G

10.  $45

    Table D

11. $44

    Table C

12.  $113

    Table E

Challenge CW11

Round and Round We Go!

Look at the table at the bottom of the page. Round the three-digit numbers to the nearest hundred and the four-digit numbers to the nearest thousand. Then find the heart that has the estimated number. Use the code in the table to color it.

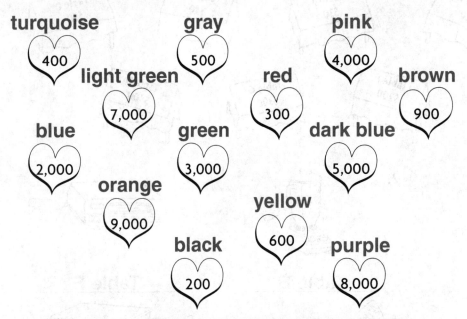

Number	Estimated Number	Color
256	300	red
1,678	2,000	blue
3,345	3,000	green
632	600	yellow
7,519	8,000	purple
8,881	9,000	orange
853	900	brown
3,780	4,000	pink
472	500	gray
187	200	black
4,574	5,000	dark blue
6,961	7,000	light green
390	400	turquoise

CW12 Challenge

Name _____

LESSON 3.1

Colorful Matches

For each sum or set of addends in Column A, find a sum or set of addends in Column B that represents the same number. Then find a sum or set of addends in Column C that represents the same number.

Draw colored lines to connect the matching sums and addends.
- Draw a red line if the sum and addends show the Commutative Property of Addition.
- Draw a blue line if the sum and addends show the Identity Property of Addition.
- Draw a green line if the sum and addends show the Associative Property of Addition.

The first one has been started for you. Trace the lines with the correct color. Red —— Blue —— Green –·–·–

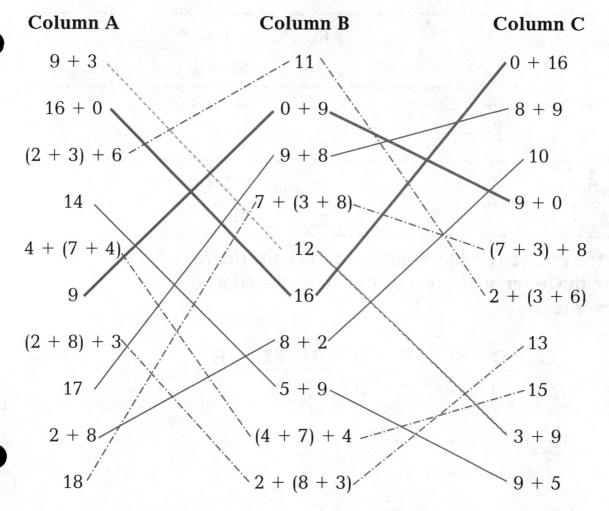

Challenge CW13

Name _____

LESSON 3.2

Missing Addend Riddle

What kind of mouse never eats cheese?

Find the missing addends.

1. 5 + __7__ = 12 E	2. __2__ + 8 = 10 P
3. 9 + __6__ = 15 A	4. 1 + __15__ = 16 T
5. __9__ + 9 = 18 M	6. 8 + __4__ = 12 S
7. 9 + __1__ = 10 R	8. __10__ + 5 = 15 C
9. __5__ + 8 = 13 O	10. __8__ + 9 = 17 M
11. __16__ + 1 = 17 U	12. 0 + __14__ = 14 O
13. __3__ + 9 = 12 U	14. __0__ + 9 = 9 E

Use the addition problems above to solve the riddle. Write the letter on the line that matches the addend below the line.

A   C   O   M   P   U   T   E   R
6   10 14 9   2   3   15 7   1

M   O   U   S   E
8   5   16 4   0

CW14   Challenge

Name _____

LESSON 3.3

Nature's Numbers

1. Mrs. Wilson bought a whale shirt, a stuffed penguin, and a bear book. About how much money did she spend?

   __about $60__

2. Mr. Jackson bought a globe and a bird feeder. About how much money did he spend?

   __about $80__

3. Ms. Garcia spent about $50. She bought a bear book and a

   __stuffed penguin, or a bird feeder__.

4. Mr. Curtis spent about $120. He bought a microscope and a

   __globe__.

5. Ms. Hunter spent about $100. She bought a whale shirt, binoculars, and a

   __stuffed penguin, or a bird feeder__.

6. Mr. Vasquez spent about $70. He bought binoculars and a

   __whale shirt__.

7. You can spend about $80 at the gift shop. You want to buy 3 gifts that are all different. You could buy:
   Possible answers are given.
   __shirt__, __book__, and __globe__ or
   __book__, __bird feeder__, and __penguin__.

Challenge CW15

Name _____

LESSON 3.4

Addition Bubbles

Use the numbers in the bubble to complete the mental math. Solve.

1. 485
   +214
   ____
     ?

   Bubble: 9, 300, 90, 2, 200, 80, 8, 10, 300

   $400 + \underline{200} = 600$

   $\underline{80} + 10 = 90$

   $5 + 4 = \underline{9}$

   | 485 |
   | +214 |
   | **699** |

2. 194
   +306
   ____
     ?

   Bubble: 2, 90, 300, 4, 0, 50, 600, 10, 100

   $\underline{100} + 300 = 400$

   $90 + \underline{0} = 90$

   $4 + 6 = \underline{10}$

   | 194 |
   | +306 |
   | **500** |

3. 686
   +212
   ____
     ?

   Bubble: 800, 14, 902, 50, 100, 700, 900, 24, 898

   $686 + \underline{14} = 700$

   $\underline{700} + 212 = 912$

   $912 - 14 = \underline{898}$

   | 686 |
   | +212 |
   | **898** |

4. 445
   +485
   ____
     ?

   Bubble: 300, 15, 930, 25, 915, 40, 20, 445, 500

   $485 + \underline{15} = 500$

   $\underline{445} + 500 = 945$

   $945 - 15 = \underline{930}$

   | 445 |
   | +485 |
   | **930** |

5. 527
   +295
   ____
     ?

   Bubble: 10, 822, 300, 5, 295, 832, 200, 827, 835

   $\underline{295} + 5 = 300$

   $527 + 300 = \underline{827}$

   $827 - \underline{5} = 822$

   | 527 |
   | +295 |
   | **822** |

CW16   Challenge

Name _____

LESSON 3.5

Palindromes

A *palindrome* is a word or phrase that reads the same forward and backward. Some examples are *Otto*, *Ada*, and *Madam, I'm Adam*.

Numbers can also be palindromes. Some examples are 88, 151, and 34,143. You can make your own number palindromes using addition. Look at the boxes.

| Choose any 2- or 3-digit number. Reverse the digits.

Add.
14
$+41$
55

55 is a palindrome. It reads the same forward and backward. | Choose any 2- or 3-digit number. Reverse the digits. 48
$+84$
Add. 132

Reverse the digits 132
of the sum. $+231$
Add. 363

The number 363 is a palindrome. |

Reverse and add until you get a palindrome.

| 1. 57
$+75$
132
$+231$
363 | 2. 153
$+351$
504
$+405$
909 | 3. 29
$+92$
121 | 4. 261
$+162$
423
$+324$
747 |

Try this out with your own 2- or 3-digit numbers. For some numbers, you need to reverse and add many times before you get a palindrome. You may need an extra piece of paper. **Check students' work. Answers will vary.**

| 5. | 6. | 7. | 8. **A 3-digit combination of 7, 8, & 9 will not work.** |

Challenge CW17

Name _____

LESSON 3.6

Get in Shape

For each pair of number sentences, each shape represents a single number. You may use the strategy *guess and check* to figure out the number that each shape represents. Write the number inside the shape.

Example:

(7) + [5] = 12

(7) − [5] = 2

1.
(9) + △6△ = 15

(9) − △6△ = 3

2.
(15) + [5] = 20

(15) − [5] = 10

3.
(80) + [20] = 100

(80) − [20] = 60

4.
(9) + △4△ = 13

(9) − △4△ = 5

5.
△4△ + [3] = 7

△4△ − 1 = [3]

6.
(9) + △7△ = 16

(9) − 2 = △7△

Answers will vary.

For Problems 7–8, there is more than one possible solution.

7.
△ + □ = 10

10 − □ = △

8.
8 − △ = ◯

8 − ◯ = △

CW18 Challenge

Name _____

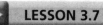 LESSON 3.7

Add Greater Numbers

Fill in the missing digits.

1. 3,47[7]
    +4,176
    ─────
    7,6[5]3

2. 2,[9]35
    +3,782
    ─────
    6,7[1]7

3. 1,63[8]
    +6,284
    ─────
    7,9[2]2

4. 4,[0]21
    +4,1[9]3
    ─────
    8,2[1]4

5. [8],516
    +1,398
    ─────
    9,[9][1]4

6. 2,82[8]
    +5,3[5]3
    ─────
    8,[1]81

7. 6,40[2]
    +4,[9]76
    ─────
    11,3[7]8

8. 4,58[8]
    +[3],219
    ─────
    7,8[0]7

9. 5,18[9]
    +2,303
    ─────
    [7],[4][9]2

10. 3,[7]3[7]
    +4,6[4]6
    ─────
    8,3[8]3

11. 2,7[9][1]
    +4,171
    ─────
    6,[9]62

12. 5,72[9]
    +4,173
    ─────
    9,[9][0]2

13. 3,[2]3[5]
    +6,371
    ─────
    9,60[0]6

14. 1,35[7]
    +[2],468
    ─────
    3,[8][2]5

15. 8,88[5]
    +4,444
    ─────
    [1][3],3[2]9

16. 5,[2]58
    +2,84[8]
    ─────
    8,1[0]6

17. 7,[8]0[2]
    +1,5[5]6
    ─────
    9,358

18. 3,82[4]
    +2,176
    ─────
    [6],[0][0]0

Challenge CW19

Name _____

LESSON 3.8

Write Expressions and Number Sentences

Write the missing number that completes the sentence. Find the code letter for each answer in the code box below. Write the code letter under each answer. Your answers will solve a riddle.

$18 + \square = 44$	$15 = \square - 12$	$\square + 48 = 62$	$120 = 20 + \square$
26	27	14	100
P	L	E	A

$\square - 45 = 45$	$56 - \square = 42$	$33 + 44 = \square$	$162 - 135 = \square$
90	14	77	27
S	E		L

$\square - 9 = 5$	$16 + 123 = \square$	$\square - 32 = 107$	$\square - 123 = 15$
14	139	139	138
E	T	T	U

$\square - 15 = 4$	$\square + 53 = 67$	$\square - 14 = 63$	$123 - \square = 23$
19	14	77	100
C	E		A

$\square - 17 = 10$	$\square + 77 = 99$	$353 = \square - 16$	$\square + 124 = 138$
27	22	369	14
L	O	N	E

```
100 = A    337 = B    19 = C     77 = (space)
14 = E     22 = O     138 = U    283 = H
75 = J     139 = T    27 = L     108 = M
369 = N    26 = P     80 = R     90 = S
```

What did the rabbits say when the farmer caught them in his garden?

Please lettuce alone.

CW20 Challenge

Name _____

LESSON 4.1

Estimate Differences

Work with a partner.

Materials:
- one number cube, with numbers 0–5
- one number cube, with numbers 1–6
- one number cube, with numbers 4–9

How to Play:
The object of the game is to get the greatest number of points.

Step 1 The first player rolls the three number cubes. This person arranges the cubes to make a 3-digit number. This person also rounds the number to the nearest hundred and writes it down.

Step 2 The second player rolls the three number cubes. This person arranges the cubes to make a 3-digit number that can be subtracted from the number the first player wrote down.

Step 3 The first player rounds that 3-digit number to the nearest hundred and subtracts it from the number that was written down.

Step 4 The answer to the estimated difference is the number of points the first player gets.

Take turns repeating Steps 1–4.

After playing the game, answer these questions.

1. What strategies did you use?

   Possible answer: When I was the first player, I tried to make my 3-digit number the greatest number possible. When I was the second player, I tried to make the number subtracted the greatest number possible.

2. What was the most difficult part of playing this game?

   Possible answer: It was hard to decide which 3-digit number to make for my partner to subtract. I had to be sure my partner would get the least number of points possible.

Challenge CW21

Name _____

LESSON 4.2

Subtraction Bubbles

Use numbers from the bubble to complete the mental math. Solve.

1. 385
    −243
    ?

    $300 - \underline{\;200\;} = 100$

    $\underline{\;80\;} - 40 = 40$

    $5 - 3 = \underline{\;2\;}$

    385
    −243
    142

2. 762
    −550
    ?

    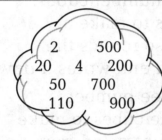

    $\underline{\;700\;} - 500 = 200$

    $60 - \underline{\;50\;} = 10$

    $2 - 0 = \underline{\;2\;}$

    762
    −550
    252

3. 835
    −250
    ?

    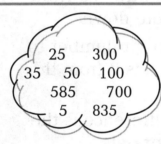

    $250 + \underline{\;50\;} = 300$

    $\underline{\;835\;} - 300 = 535$

    $535 + 50 = \underline{\;585\;}$

    835
    −250
    585

4. 298
    −172
    ?

    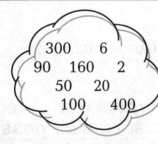

    $200 - \underline{\;100\;} = 100$

    $\underline{\;90\;} - 70 = 20$

    $8 - 2 = \underline{\;6\;}$

    298
    −172
    126

5. 618
    −495
    ?

    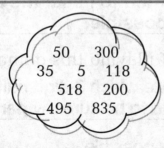

    $\underline{\;495\;} + 5 = 500$

    $618 - 500 = \underline{\;118\;}$

    $118 + \underline{\;5\;} = 123$

    618
    −495
    123

CW22 Challenge

Subtract 3- and 4-Digit Numbers

Subtract. Connect the dots in order from the least difference to the greatest difference.

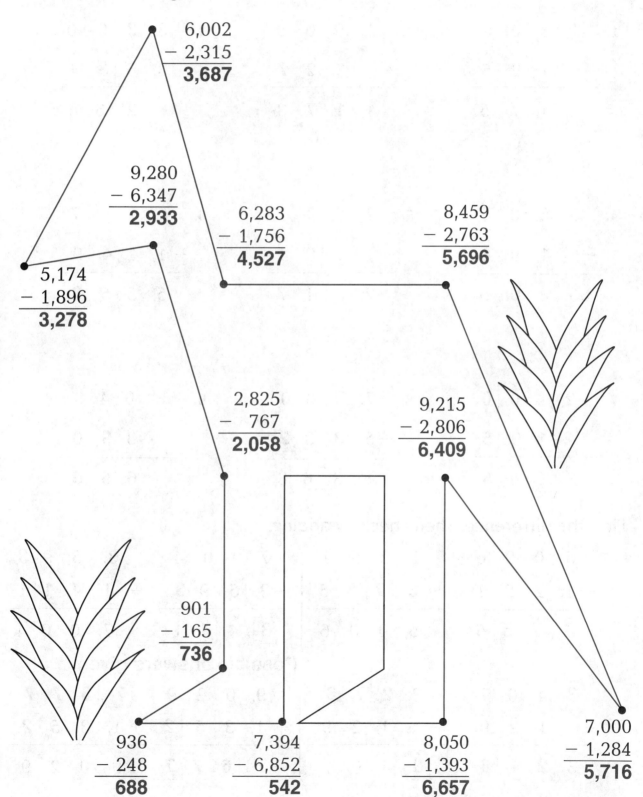

Challenge CW23

Name _____

Missing Digits

Fill in the missing digits.

1. 1,30**0**
    − 125
    ———
    1,1**7**5

2. 1,500
    − 32**7**
    ———
    1,**1**73

3. 3,**9**00
    −3,6**3**4
    ———
       266

4. 5,50**3**
    − 198
    ———
    5,3**0**5

5. 2,6**0**1
    − 284
    ———
    2,31**7**

6. 6,707
    −1,1**4**9
    ———
    5,55**8**

7. 2,40**0**
    −**2**,195
    ———
       2 05

8. 7,**7**00
    −5,33**2**
    ———
    2,368

9. **1**,040
    − 3**9**0
    ———
       65**0**

Find the difference. Then check by adding.

   5,006   1,251   7,004   3,59**3**
 − 1,251   +**3**,**7 5**5   − 3,59**3**   +**3**,**4 1**1
 ———————   ——————   ———————   ———————
 3,**7 5**5   5,006   **3**,**4 1**1   7,004

                                    Possible answers given.

   3,407   3,**2 7 8**   9,029   **7**,**6 7 7**
 −   129   +   129   − 1,35**2**   +1,35**2**
 ———————   ——————   ———————   ———————
 3,**2 7 8**   **3**,407   **7**,**6 7 7**   9,029

CW24 Challenge

Name _____

LESSON 4.5

Planning a Party

Mrs. Laff is catering a party for 14 girls and 14 boys. Use the chart to help her plan the party.

Cupcakes	16 in a box
Pizza	1 pizza has 8 slices
Lemonade	1 bottle fills 10 glasses
Chips	9 servings in a bag
Balloons	12 in a bag

1. Each person will get 1 cupcake. How many boxes of cupcakes are needed?

    2 boxes

2. How many cupcakes will be left if each person eats one cupcake?

    4

3. How many pizzas would give each person 2 slices?

    7 pizzas

4. How many bottles of lemonade would give each person 2 glasses of lemonade?

    6 bottles of lemonade

5. Mrs. Laff bought 3 bags of chips. About how many servings per person is this?

    about 1 serving

6. Mrs. Laff plans to make a name card for each person. How many name cards must she make?

    28

7. Mrs. Laff plans to decorate the party room with balloons. She wants to use at least 50 balloons. How many bags should she buy?

    5 bags

8. A package of party napkins costs $8.85. Mrs. Laff pays for the napkins with $10. How much change will she get?

    $1.15

9. Mrs. Laff has $120 to spend on food for this party. The cupcakes will cost $32.85. The pizza will cost $48.00. The lemonade will cost $8.19. The chips will cost $10.94. Will she need more money for the food? Explain.

    No; because $30 + $40 + $8 + $10 = $88, which is less than $120.

10. Mrs. Laff can arrange the tables in 3 different ways. Each way seats a different number of people. She is able to seat 20, 25, or 36. Which way should she arrange her tables? Explain.

    Since she needs seats for 28, she will have to arrange the tables to seat 36.

Challenge CW25

Colorful Sets

Use this key to color the design below.

Section A and all sections with equivalent sets of money	Red
Section B and all sections with equivalent sets of money	Yellow
Section C and all sections with equivalent sets of money	Green
Section D and all sections with equivalent sets of money	Blue

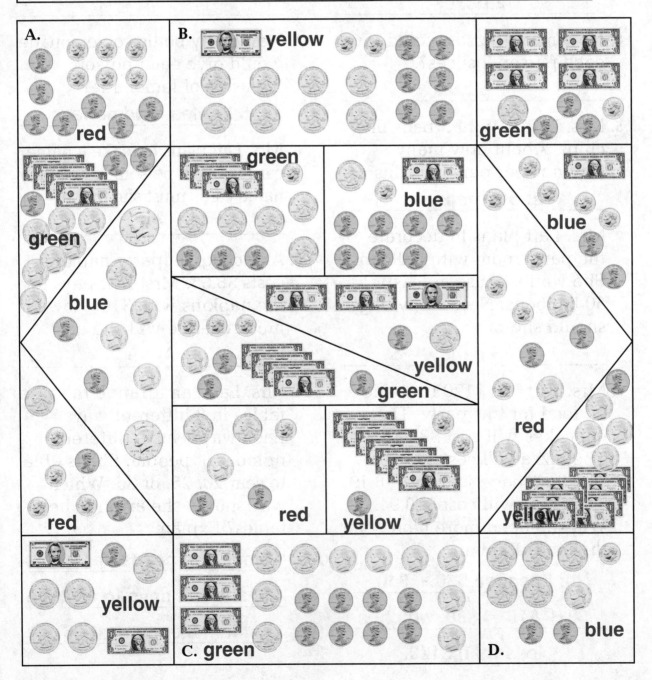

CW26 Challenge

Name _____

LESSON 5.2

Paying Cash

Use the table for 1–5. *Act it out* to solve.
Check students' models. Possible answers are given.

1. Kelly wants to buy a cheeseburger and a fruit cup for lunch. She has 5 $1 bills, 4 quarters, and 8 nickels. What set of bills and coins can she use?

   4 $1 bills, 2 quarters, 1 dime, 1 nickel

2. Jason is buying a chef salad and a lemonade. He has 1 $5 bill, 2 $1 bills, 8 quarters, 3 dimes, and 2 nickels. What set of bills and coins can he use?

   1 $5 bill, 3 quarters, 1 dime, 1 nickel

3. Mizra has 3 $1 bills, 5 quarters, and 3 dimes. She wants to buy a cheeseburger and chips. What set of bills and coins can she use?

   3 $1 bills, 4 quarters

4. Evan has 4 $1 bills, 3 quarters, 3 dimes, and 5 pennies. He's buying a chef salad and wants to keep 2 quarters to play a video game later. What set of bills and coins should he use?

   4 $1 bills, 1 quarter, 3 dimes

5. Sierra has 2 $1 bills, 4 quarters, 8 dimes, and 3 nickels. Describe two different, equivalent sets of bills and coins she could use to buy a cheeseburger.

   2 $1 bills, 4 quarters, 2 dimes, and 1 nickel; 2 $1 bills, 4 quarters, 1 dime, and 3 nickels

Challenge   CW27

Shopping at the Pet Store

$1.49 $0.46 $0.89 $0.33 $2.98

1. Jim has

Does Jim have enough money to buy a brush?

_____**no**_____

2. Carrie has

Does Carrie have enough money to buy 2 cans of cat food?

_____**yes**_____

3. Lisa has

How much more money does Lisa need to buy a bowl?

_____**$0.14**_____

4. Fred has

How much money will Fred have left if he buys a can of cat food?

_____**$0.25**_____

5. Jane pays for a toy mouse with 8 coins. What coins does she use?

3 quarters, 1 dime, 4 pennies

6. Paul pays for a ball with 4 coins. What coins does he use?

1 quarter, 2 dimes, 1 penny

Name _____

Make Change

Complete the table of items needed to make a cake.

Paid	Cost of Item	Change Received
8 quarters	$0.74	$0.06
2 one-dollar bills	$1.29	$0.71
Possible answer: $5.00	$1.49	$3.51
Possible answer: $1.00	$0.65	$0.35
$1 bill + 1 quarter	$1.05	$0.20
4 quarters + 1 dime	$1.09	$0.01
2 one-dollar bills + 2 quarters	$2.35	$0.15

Challenge   CW29

Name _____

LESSON 5.5

Money Matters

Write the missing numbers.

1. $ 2 . 5 5
    + 1 . 9 8
    ─────────
    $ [4] . 5 [3]

2. $ 2 . 9 5
    + 3 . [6] 9
    ─────────
    $ 6 . 6 4

3. $ 4 . 5 9
    + [2] . 3 [9]
    ─────────
    $ 6 . 9 8

4. $ 2 3 . [9] 1
    + [1] 5 . 3 9
    ─────────
    $ 3 9 . 3 0

5. $ 3 . 9 5
    − 1 . 4 9
    ─────────
    $ [2] . 4 [6]

6. $ 4 . 5 0
    − 1 . [2] 8
    ─────────
    $ 3 . 2 2

7. $ 6 . 5 9
    − [1] . 9 [2]
    ─────────
    $ 4 . 6 7

8. $ 4 0 . [7] 5
    − 2 [1] . 4 9
    ─────────
    $ 1 9 . 2 6

For problems 9–12, use the table.

Sandwiches	
Ham	$3.89
Cheese	$2.35
Chicken	$3.19
Peanut butter and jelly	$1.65

9. Bob buys 2 peanut butter and jelly sandwiches. Should he give the clerk a $1 bill, a $5 bill, or a $10 bill? Explain.

   Possible answer: a $5 bill because the cost is between $3.00 and $4.00

10. Joan buys a sandwich. She gives the clerk a $5 bill. Her change is $2.65. What kind of sandwich does she buy?

    cheese

11. Mr. Riley buys one of each kind of sandwich. How much should he give the clerk?

    $11.08

12. Make up your own problem about sandwiches you will buy for yourself and a friend. Write the problem so that the solver must use estimation, addition, and subtraction. Have a classmate solve it.

    Check students' problems. Sample answer: I have $6.00 to buy 2 sandwiches, one for me and one for my friend Ben. Which sandwiches can I buy? How much change will I get?

CW30 Challenge

Name _____

LESSON 6.1

Find the Time

Read the time on each clock. For each time, draw a line to connect the words and numbers that you used. The words and numbers may go up, down, left, right, or diagonal. The first one is done for you.

1.
2.
3.
4.

5.
6.
7.
8.

25 minutes	eight	before	17 minutes	quarter
after	after	ten	past	past
two	eight	three	before	seven
before	35 minutes	quarter	nine	13 minutes
half	to	two	five	8 minutes
one	past	before	after	before
half	52 minutes	seven	6 minutes	eleven

Challenge CW31

Time for a Riddle

To solve the riddle, match the letter in the circle and the time in the box below. Write the letter on the line above the box.

1. At 15 minutes after nine, I start school. (l)
2. At four thirty I had soccer practice. (a)
3. I get out of school when it is 30 minutes after three. (f)
4. I have music class at 30 minutes after two. (b)
5. I eat dinner at 15 minutes before six. (s)
6. At six fifteen my alarm wakes me up. (a)
7. We looked up at the moon at eight forty-five. (u)
8. I have a doctor's appointment at 30 minutes after ten. (o)
9. By twelve forty-five we had the lunch dishes put away. (n)
10. When it is ten thirty, my mom is asleep. (n)
11. I had to get ready for bed at 30 minutes after eight. (c)
12. We looked for a star at ten forty-five. (s)
13. When it was 15 minutes before four, I played baseball with my friends. (e)
14. When it was 15 minutes after one, I awoke to the sound of thunder. (d)
15. At three fifteen I will visit my friend. (t)
16. I have never stayed awake later than eleven thirty. (e)
17. The sun will rise tomorrow at five fifteen. (g)
18. When I was sick, I slept until 15 minutes after eleven. (h)

What has a __f__ __a__ __c__ __e__ and __h__ __a__ __n__ __d__ __s__

| 3:30 P.M. | 6:15 A.M. | 8:30 P.M. | 3:45 P.M. | | 11:15 A.M. | 4:30 P.M. | 10:30 P.M. | 1:15 A.M. | 5:45 P.M. |

__b__ __u__ __t__ __n__ __o__ __l__ __e__ __g__ __s__ ?

| 2:30 P.M. | 8:45 P.M. | 3:15 P.M. | | 12:45 P.M. | 10:30 A.M. | | 9:15 A.M. | 11:30 P.M. | 5:15 A.M. | 10:45 P.M. |

Solve the riddle. __a clock__

CW32 Challenge

Name _____

Time Flies

What do pilot rabbits fly?

Use the clocks to answer the riddle. Find the clock that matches each time written at the bottom of the page. Write the letter of the clock in the box above the time.

H	A	R	E
15 minutes after 9:15	1 hour after 7:30	30 minutes after 4:15	1 hour 15 minutes after 10:00

P	L	A	N	E	S
15 minutes after 9:45	30 minutes after 11:45	15 minutes after 8:15	45 minutes after 1:30	30 minutes after 10:45	45 minutes after 3:15

LESSON 6.3

Challenge CW33

Name _____

LESSON 6.4

Problem Solving Skill
Make a Schedule

Mr. Frank's class is going to a nature center. Mr. Frank drew these clocks to show when each activity begins. Each activity ends just as the next activity begins.

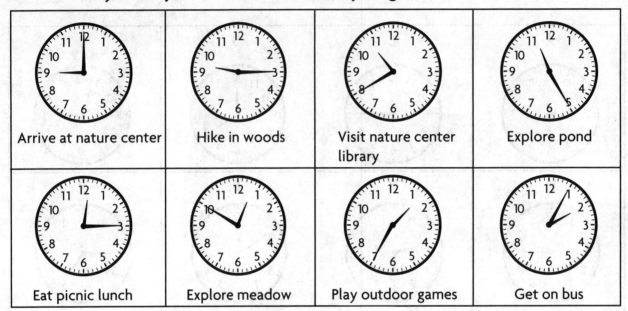

1. Complete the schedule.

Activity	Time	Elapsed Time
Arrive	9:00 A.M.–9:15 A.M.	15 minutes
Hike	9:15 A.M.–10:40 A.M.	1 hour 25 minutes
Visit library	10:40 A.M.–11:25 A.M.	45 minutes
Explore pond	11:25 A.M.–12:15 P.M.	50 minutes
Eat lunch	12:15 P.M.–12:50 P.M.	35 minutes
Explore meadow	12:50 P.M.–1:35 P.M.	45 minutes
Play games	1:35 P.M.–2:05 P.M.	30 minutes

2. Which activity lasts the longest? _____hike_____

3. How much time in all will the class spend at the nature center? ___5 hours 5 minutes___

CW34 Challenge

Name _____

LESSON 7.1

Multiply in the Sky

Isabel is on her first airplane flight. She looks out the window and writes down what she sees. Complete each number sentence. Then draw a picture of what Isabel sees. **Possible answers are given. Check students' drawings.**

1. $3 \times 3 =$ __9 houses__
2. $2 \times 2 =$ __4 trucks__
3. $4 \times 5 =$ __20 trees__
4. $2 \times 4 =$ __8 cars__
5. $3 \times 2 =$ __6 swing sets__

Challenge CW35

Pattern Plot

Write a multiplication fact using 2 or 5 that will give each product. The first one is done for you. Then color in the boxes with facts using 2. What do you see?

8 _4_ × _2_	14 _7_ × _2_	12 _6_ × _2_	24 _12_ × _2_
15 _3_ × _5_	35 _7_ × _5_	55 _11_ × _5_	22 _11_ × _2_
12 _6_ × _2_	16 _8_ × _2_	4 _2_ × _2_	6 _3_ × _2_
22 _11_ × _2_	5 _1_ × _5_	25 _5_ × _5_	45 _9_ × _5_
4 _2_ × _2_	18 _9_ × _2_	14 _7_ × _2_	16 _8_ × _2_

I see a _2_!

CW36 Challenge

Name _____

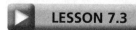
LESSON 7.3

Dance on Arrays

You have been asked to complete a design for a dance floor using arrays of colors. You can use as much red, green, blue, yellow, and orange as you want. Make a list of the colors and arrays you use. The design has been started for you in black. **Designs will vary. Check to see that students have made accurate multiplication sentences for the arrays they colored.**

2 × 2 = __4__ black squares   ____ × ____ = _____

____ × ____ = _____   ____ × ____ = _____

____ × ____ = _____   ____ × ____ = _____

Challenge   CW37

Puzzling Products

Find the product in Column 2 for each problem in Column 1. Then write the letter of the product on the line in front of the problem.

Column 1

- __N__ 1. 9 × 3
- __S__ 2. 5 × 5
- __B__ 3. 2 × 4
- __Y__ 4. 7 × 3
- __E__ 5. 3 × 5
- __U__ 6. 8 × 2
- __A__ 7. 6 × 3
- __M__ 8. 2 × 5
- __D__ 9. 3 × 8
- __R__ 10. 5 × 6
- __O__ 11. 2 × 3

Column 2

- (A) 18
- (B) 8
- (D) 24
- (E) 15
- (M) 10
- (N) 27
- (O) 6
- (R) 30
- (S) 25
- (U) 16
- (Y) 21

Use your answers to decode the sentence below. The problem number is under each line. Write the letter of your answer from Column 1 on the line.

A	D	D	E	N	D	S		A	R	E
7.	9.	9.	5.	1.	9.	2.		7.	10.	5.

N	U	M	B	E	R	S		Y	O	U		A	D	D
1.	6.	8.	3.	5.	10.	2.		4.	11.	6.		7.	9.	9.

CW38 Challenge

Name _____

LESSON 7.5

What's the Question?

For 1–6, complete the problem with a question so that the answer given is correct. **Possible questions are given.**

1. Four friends walked 4 blocks to school. Each one is carrying 3 books.

   How many books are there?

   Too much information;
   12 books

2. Doug bought 6 pencils. They were on sale today for 5¢ less than they were yesterday.

   How much did he spend on pencils?

   Not enough information

3. Xavier bought 4 ice cream cones. The clerk gave him $0.50 in change.

   How much did he spend on ice-cream cones?

   Not enough information

4. Misty's piggy bank has 3 times as many nickels as dimes. She has 3 more dimes than quarters. Misty has 8 dimes.

   How many nickels does she have?

   Too much information;
   24 nickels

5. When all of Pam's cousins come to visit, the number of children in her house will double. Pam has 2 cousins.

   How many children will be in her house?

   Right amount of information;
   4 children

6. A three-digit number has a ones digit that is 4 times its tens digit. The tens digit is 2. The hundreds digit is 1 less than the tens digit.

   What is the ones digit?

   Too much information; 8

Challenge   CW39

Name _____

LESSON 8.1

Fit Feasting Facts

MyPyramid is a guide to healthy eating with exercise. There are different MyPyramids for different people. This pyramid shows how a third-grader can build a healthy diet by eating different kinds of foods.

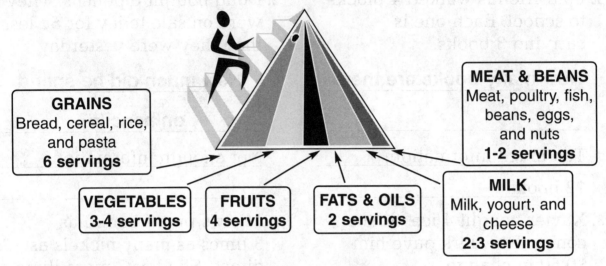

Use MyPyramid to answer the questions below.

1. How many servings of fruit should you eat every day?

   4 servings

2. How many servings of grains should you eat every day?

   6 servings

3. Every day, Max eats the most servings of vegetables recommended by MyPyramid. How many servings of vegetables does Max eat in a week?

   28 servings

4. Every day, Julie eats the fewest servings of vegetables recommended by MyPyramid. How many servings from this group does Julie eat in 3 days?

   9 servings

5. Every day, Juan eats the least number of servings from the milk, yogurt, and cheese group shown in the pyramid. How many servings of this group does Juan eat in 5 days?

   10 servings

6. Tamala eats the greatest number of servings of the meat, poultry, fish, beans, eggs, and nuts group every day. How many servings of this group does she eat in 8 days?

   16 servings

CW40 Challenge

Pondering Products

Find the product for each Column 1 problem in Column 2. Then write the product's circled letter on the line in front of the problem.

	Column 1		Column 2
__I__	1. 4×9	Ⓔ	8
__T__	2. 3×2	Ⓕ	0
__O__	3. 2×8	Ⓘ	36
__F__	4. 0×4	Ⓧ	45
__R__	5. 8×4	Ⓜ	40
__U__	6. 5×7	Ⓝ	20
__E__	7. 4×2	Ⓞ	16
__S__	8. 2×7	Ⓡ	32
__X__	9. 5×9	Ⓢ	14
__M__	10. 8×5	Ⓣ	6
__N__	11. 5×4	Ⓤ	35

Use your answers to decode the sentence below. The problem number under each blank tells you where to look in Column 1. Write the letter of your answer from Column 1 on the blank.

__F__ __O__ __U__ __R__   __T__ __I__ __M__ __E__ __S__   __F__ __O__ __U__ __R__
4. 3. 6. 5.   2. 1. 10. 7. 8.   4. 3. 6. 5.

__I__ __S__   __S__ __I__ __X__ __T__ __E__ __E__ __N__ .
1. 8.   8. 1. 9. 2. 7. 7. 11.

Name _____

LESSON 8.3

Problem Solving Strategy

Look for a Pattern

Play with a partner. **Check students' work**

Materials: three number cubes: one numbered 1–6, one numbered 7–12, and one numbered 13–18

How to Play:

Players take turns as rollers and pattern makers. The roller rolls the three number cubes and writes the three numbers in order from least to greatest on a line below under his or her partner's column of patterns.

The pattern maker finds a rule and continues the pattern for four more numbers. For example, for the numbers 5, 10, and 18, the rule could be (+ 5, + 8) or (× 2, + 8).

The seventh number in the pattern is the number of points the pattern maker scores. Find the sum of each player's seventh numbers. The winner is the player with the greater sum.

Player 1 Patterns	**Player 2 Patterns**
___, ___, ___, ___, ___, ___, ___	___, ___, ___, ___, ___, ___, ___
___, ___, ___, ___, ___, ___, ___	___, ___, ___, ___, ___, ___, ___
___, ___, ___, ___, ___, ___, ___	___, ___, ___, ___, ___, ___, ___
___, ___, ___, ___, ___, ___, ___	___, ___, ___, ___, ___, ___, ___
___, ___, ___, ___, ___, ___, ___	___, ___, ___, ___, ___, ___, ___
___, ___, ___, ___, ___, ___, ___	___, ___, ___, ___, ___, ___, ___

Player 1: Find the sum of the last numbers in each of your patterns.

Player 2: Find the sum of the last numbers in each of your patterns.

_____

CW42 Challenge

Name _____

The Factor Game

Play with a partner.

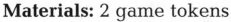

Materials: 2 game tokens
30 number cards including ten 7s, ten 8s, and ten 9s

How to Play:
Place the game tokens on Start. Shuffle the number cards. Turn the number cards upside down or place them in a bag.

Take turns drawing a number card. Move to the closest space with a missing factor that matches the number on the card. Write the factor in the space. The first player to reach Finish wins!

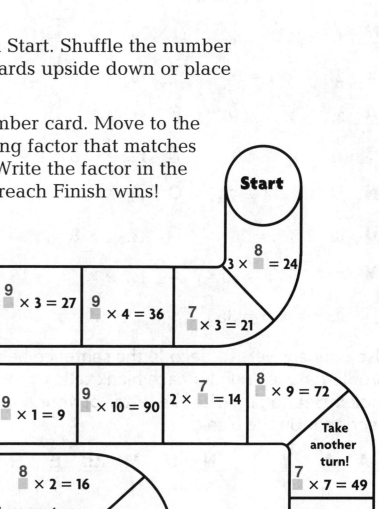

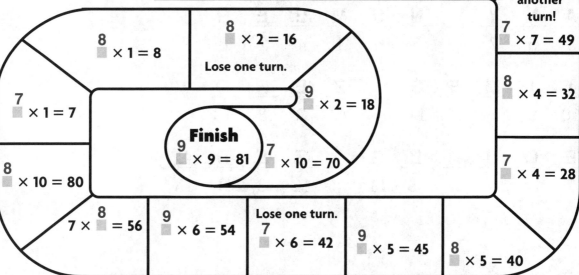

Challenge CW43

Find Those Factors

Find the missing number for each Section 1 problem in Section 2. Then write the number's circled letter on the line in front of the problem.

Section 1

__M__ 1. $2 \times \blacksquare = 8$
__E__ 2. $\blacksquare \times 1 = 9$
__A__ 3. $\blacksquare \times 5 = 40$
__B__ 4. $8 \times \blacksquare = 0$
__N__ 5. $\blacksquare \times 4 = 24$
__U__ 6. $2 \times 6 = \blacksquare$
__Y__ 7. $\blacksquare \times 4 = 20$
__R__ 8. $\blacksquare \times 5 = 35$

__I__ 9. $1 \times \blacksquare = 1$
__T__ 10. $7 \times \blacksquare = 21$
__S__ 11. $4 \times 7 = \blacksquare$
__Z__ 12. $2 \times 8 = \blacksquare$
__O__ 13. $5 \times 7 = \blacksquare$
__Q__ 14. $4 \times 8 = \blacksquare$
__L__ 15. $9 \times \blacksquare = 18$

Section 2

Ⓝ 6   Ⓔ 9
Ⓐ 8   Ⓘ 1
Ⓛ 2   Ⓠ 32
Ⓑ 0   Ⓞ 35
Ⓨ 5   Ⓜ 4
Ⓣ 3   Ⓩ 16
Ⓢ 28 Ⓡ 7
Ⓤ 12

Use your answers to decode the sentence below. The problem number under each blank tells you where to look in Section 1. Write the letter of your answer from Section 1 on the blank.

__A__ __N__ __Y__   __N__ __U__ __M__ __B__ __E__ __R__
 3   5   7   5   6   1   4   2   8

__T__ __I__ __M__ __E__ __S__   __Z__ __E__ __R__ __O__
10   9   1   2   11   12   2   8   13

__E__ __Q__ __U__ __A__ __L__ __S__   __Z__ __E__ __R__ __O__
 2 14   6   3 15 11   12   2   8   13

CW44 Challenge

Name _____

LESSON 9.1

The Array Game

Play alone or with a partner.

Materials: 10 × 10 grid for each player, two number cubes labeled 1–6, crayons or colored pencils

How to Play:

- Roll the number cubes. If you are playing with a partner, take turns rolling.
- Shade an array on your grid with a length and width that correspond to the numbers you rolled.

  Example: Suppose you roll and 6 .

  Shade an array that is 2 squares wide and 6 squares long or 6 squares wide and 2 squares long. You may place the array anywhere on your grid. Arrays cannot overlap.

- The object of the game is to shade as much of the grid as possible during the time that you have to play the game.

Score: Your score is the total number of squares that you have shaded when time runs out. **Check students' work.**

Game 1

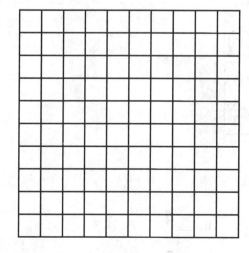

Score _____

Game 2

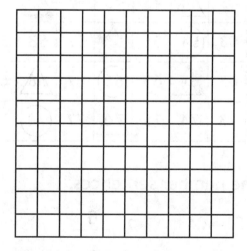

Score _____

Challenge CW45

LESSON 9.2

Name _____

Number Patterns

A **multiple** is a number that is the product of a given number and another whole number. Some of the multiples of 3 are: 3, 6, 9, and 12.

1. On the number chart below, put a triangle around the numbers that are multiples of 4.
   Check students' charts.
2. Circle all the numbers that are multiples of 6.
   Check students' charts.
3. Shade all the numbers that are multiples of 8.
   Check students' charts.
4. List the numbers that have triangles around them and are also circled and shaded. __**24, 48, 72**__
5. Are there any shaded numbers that do not have triangles around them? __**no**__

1	2	3	4	5	6	7	8	9	10
11	12	13	14	15	16	17	18	19	20
21	22	23	24	25	26	27	28	29	30
31	32	33	34	35	36	37	38	39	40
41	42	43	44	45	46	47	48	49	50
51	52	53	54	55	56	57	58	59	60
61	62	63	64	65	66	67	68	69	70
71	72	73	74	75	76	77	78	79	80

Complete the number sentences.

6. __3__ × 8 = 24
7. __6__ × 8 = 48
8. __4__ × 8 = 32
9. __4__ × 6 = 24
10. __8__ × 6 = 48
11. __5__ × 6 = 30

CW46 Challenge

Name _____

LESSON 9.3

Square Time

A **square array** is an array that is the same number of squares long as it is wide.

Complete the table, listing the sizes and products of some square arrays.

Square Arrays		
Length	Width	Total Squares
1	1	1
2	2	4
3	3	9
4	4	16
5	5	25
6	6	36
7	7	49
8	8	64

Use the table to solve. **See students' work.**

1. Jay and Barb each made a square array. Barb used more squares than Jay. Together they used 100 squares. How big was each array?

   Jay's was 6 × 6; Barb's was 8 × 8.

2. Tim, Mark, Dave, and Paul each made a square array. Together they used 100 squares. How big was each array?

   Possible answer: Each made a 5 × 5 square.

3. Sharon, Gayle, Joy, and Bev each made a square array. Joy and Bev used 22 more squares than Sharon and Gayle. Altogether the 4 girls used 122 squares. How big was each array?

   Joy and Bev each made a 6 × 6 array; Sharon and Gayle each made a 5 × 5 array.

Challenge CW47

Finding Factor Pairs

What kind of fruit is always grumpy?

To find out, draw a line to match each clue to the correct factor pair. Write the factor pair's code letter above the clue number at the bottom of page.

Factor Pairs	Code Letter
4,6	A
6,7	P
4,8	S
5,6	C
4,7	A
7,9	R
7,8	L
3,7	E
6,6	P
6,8	B

1. Their product is equal to 15 + 15.
2. Their product is odd. Their difference is 2.
3. Their product is equal to 3 × 8.
4. Their product is between 40 and 50. Their sum is even.
5. Their product is equal to 14 + 14.
6. Their product is about 40. Their difference is 1.
7. Their product is between 35 and 40.
8. Their product is greater than 50. Their difference is 1.
9. Their product is equal to 28 − 7.
10. Their product is even. Their difference is 4.

```
 C   R   A   B      A   P   P   L   E   S
 1.  2.  3.  4.     5.  6.  7.  8.  9.  10.
```

CW48 Challenge

LESSON 9.5

Name _____

Row after Row

Sam displays apples in rows in his supermarket.

- Circle two fact arrays in each display of apples.
- Write the facts you used to find the total number of apples. **Check students' arrays. Possible answers are given.**

1.

   $3 \times 4 = 12; 3 \times 8 = 24;$

   $12 + 24 = 36;$ there are

   36 apples.

2.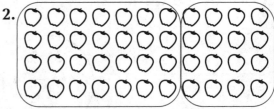

   $4 \times 7 = 28; 4 \times 4 = 16;$

   $28 + 16 = 44;$ there are

   44 apples.

3.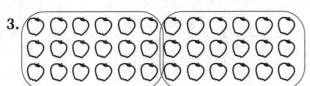

   $3 \times 6 = 18; 3 \times 6 = 18;$

   $18 + 18 = 36;$ there are

   36 apples.

4.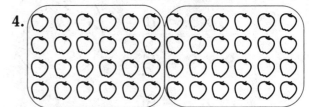

   $4 \times 6 = 24; 4 \times 6 = 24;$

   $24 + 24 = 48;$ there are

   48 apples.

5.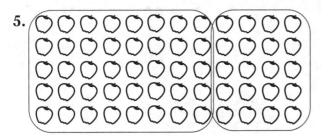

   $5 \times 8 = 40; 5 \times 4 = 20;$

   $40 + 20 = 60;$ there are

   60 apples.

6.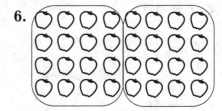

   $4 \times 4 = 16; 4 \times 4 = 16;$

   $16 + 16 = 32;$ there are

   32 apples.

Challenge CW49

Name _____

LESSON 10.1

Combination Challenge

Use the numbers and symbols in each circle to make 4 different number sentences. Each number and symbol can be used only once in a sentence, but they can be used in more than one sentence. The first problem is done for you. **Possible answers are given. Check students' sentences.**

1. $3 \times 9 = 27$
   $3 \times 4 = 12$
   $3 \times 7 = 21$
   $2 \times 7 = 14$

   (Circle: 3 9 1 × = 4 2 7)

2. $0 \times 4 = 0$
   $4 \times 4 = 16$
   $10 \times 4 = 40$
   $1 \times 4 = 4$

   (Circle: 0 4 1 × = 0 4 6)

3. $1 \times 9 = 9$
   $0 \times 9 = 0$
   $9 \times 9 = 81$
   $10 \times 9 = 90$

   (Circle: 9 0 1 × = 0 8 9)

4. $4 \times 9 = 36$
   $7 \times 9 = 63$
   $7 \times 7 = 49$
   $6 \times 9 = 54$

   (Circle: 7 5 9 6 × = 4 3 7)

5. $2 \times 9 = 18$
   $8 \times 9 = 72$
   $7 \times 2 = 14$
   $8 \times 2 = 16$

   (Circle: 2 9 1 × = 8 4 6 7)

6. $7 \times 9 = 63$
   $6 \times 9 = 54$
   $8 \times 7 = 56$
   $4 \times 9 = 36$

   (Circle: 7 9 6 × = 4 5 8 3)

7. $5 \times 9 = 45$
   $10 \times 4 = 40$
   $10 \times 5 = 50$
   $5 \times 1 = 5$

   (Circle: 5 0 9 1 × = 4 5 0 4)

8. $8 \times 8 = 64$
   $8 \times 6 = 48$
   $10 \times 6 = 60$
   $10 \times 8 = 80$

   (Circle: 0 8 1 6 × = 6 8 4 0)

CW50 Challenge

Name _____

LESSON 10.2

What's the Rule?

Ken and Mary are playing a game. First Mary draws a design and then Ken thinks of a rule and draws his design. When the table is completed, Mary tries to guess the rule Ken followed. They change roles and play again. Help Mary and Ken find the rules. **Designs may vary. Check rules.**

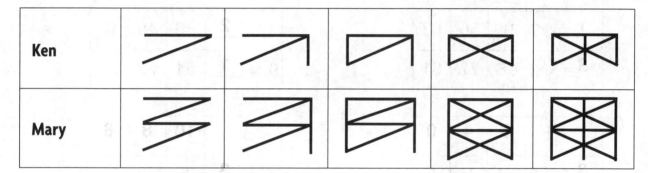

Rule: **Multiply the number of squares in Mary's design by 3.**

Rule: **Multiply the number of lines in Ken's design by 2.**

Choose a partner and play Mary and Ken's game. **Designs will vary.**

Your Name _____					
Your partner's name _____					

Rule: **Answers will vary.**

Challenge CW51

Missing Factors

Here are some Multiplication Squares to challenge your multiplication skills! Fill in the missing factors to complete the squares. **Possible answers are given.**

1.

×	3	5	9
2 × 4	24	40	72
3 × 3	27	45	81
5 × 2	30	50	90

2.

×	4	10	7
2 × **3**	24	60	42
3 × **3**	36	90	63
8 × **0**	0	0	0

3.

×	2	8	9
2 × 5	20	80	90
1 × 7	14	56	63
3 × 3	18	72	81

4.

×	6	8	3
2 × **2**	24	32	12
3 × **2**	36	48	18
9 × **1**	54	72	27

5.

×	6	4	0
3 × 3	54	36	0
7 × 1	42	28	0
2 × 2	24	16	0

6.

×	10	8	6
3 × **3**	90	72	54
1 × **5**	50	40	30
4 × **2**	80	64	48

7.

×	2	4	5
9 × 1	18	36	45
1 × 7	14	28	35
2 × 5	20	40	50

8.

×	2	4	6
2 × **2**	8	16	24
1 × **5**	10	20	30
2 × **3**	12	24	36

Name _____

LESSON 10.4

Property Match Game

Play with a partner.

Materials: expression cards shown below; scissors

How to Play:
- Cut apart the expression cards and place them facedown on a table.
- Players take turns. Turn over two cards. Determine whether the cards are an example of a multiplication property. If so, name the property. If not, place the cards back on the table facedown.
- If the property is named correctly, keep the cards. If not, place the cards back on the table facedown.
- When all the cards have been picked up, the player with more cards wins the game!

5×6	$(2 \times 2) + (2 \times 7)$	7
2×9	0×7	$8 \times (4 \times 2)$
9×1	$(7 \times 2) \times 5$	$(3 \times 2) \times 4$
0	$(4 \times 5) + (4 \times 3)$	6×5
$(8 \times 4) \times 2$	$7 \times (2 \times 5)$	7×1
$3 \times (2 \times 4)$	1×9	4×8

Make up your own set of cards. Trade with another pair of classmates, and play again.

Challenge CW53

Special Delivery

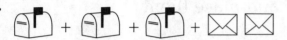

In each problem, the mailboxes have the same number of letters inside. Write the total number of letters for each problem.

1. Key: 🏠 = 5 letters
   17 letters

2. Key: 🏠 = 10 letters
   34 letters

3. Key: 🏠 = 8 letters
   39 letters

4. Key: 🏠 = 9 letters
   25 letters

5. Key: 🏠 = 7 letters
   20 letters

6. Key: 🏠 = 9 letters
   51 letters

7. 4 × 🏠 + ✉✉
   Key: 🏠 = 3 letters
   14 letters

8. 3 × 🏠 + ✉✉✉✉
   Key: 🏠 = 8 letters
   28 letters

9. 6 × 🏠 − ✉✉✉
   Key: 🏠 = 5 letters
   27 letters

10. 8 × 🏠 + ✉✉✉✉✉
    Key: 🏠 = 9 letters
    77 letters

CW54 Challenge

Name _____   LESSON 11.1

Paintbrush Division

The jars in each row need to be filled with the same number of paintbrushes. Draw the paintbrushes in each jar and complete the number sentence. **Check students' drawings.**

1.

   Total number of paintbrushes: 12

   Paintbrushes in each jar: __4__

   12 ÷ 3 = __4__

2.

   Total number of paintbrushes: 8

   Paintbrushes in each jar: __2__

   8 ÷ 4 = __2__

3.

   Total number of paintbrushes: 12

   Paintbrushes in each jar: __6__

   12 ÷ 2 = __6__

4.

   Total number of paintbrushes: 15

   Paintbrushes in each jar: __3__

   15 ÷ 5 = __3__

5.

   Total number of paintbrushes: 18

   Paintbrushes in each jar: __6__

   18 ÷ 3 = __6__

6.

   Total number of paintbrushes: 20

   Paintbrushes in each jar: __5__

   20 ÷ 4 = __5__

Complete the chart.

	Number of Paintbrushes	Number of Jars	Number of Paintbrushes in Each Jar
7.	24	4	6
8.	21	3	7
9.	30	5	6

Challenge CW55

LESSON 11.2

Name _____

Animal Division

Separate the animals into groups. Draw a circle around each group. Then complete the number sentence. **Check students' work.**

1. 4 cats in each group

8 ÷ 4 = __2__

2. 3 dogs in each group

9 ÷ 3 = __3__

3. 3 birds in each group

12 ÷ 3 = __4__

4. 5 turtles in each group

10 ÷ 5 = __2__

5. 5 mice in each group

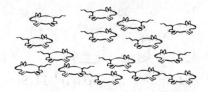

15 ÷ 5 = __3__

6. 3 fish in each group

18 ÷ 3 = __6__

Complete the chart.

	Number of Animals	Number in Each Group	Number of Equal Groups
7.	18 puppies	3	6
8.	20 kittens	4	5
9.	24 gerbils	6	4
10.	30 guinea pigs	5	6

CW56 Challenge

Name _____

LESSON 11.3

Missing Numbers

Complete each table.

1.

Number of Students	Number of Hands
1	2
4	**8**
6	**12**
2	**4**
9	18
7	14
5	10

2.

Number of Tricycles	Number of Wheels
1	3
3	**9**
7	**21**
6	**18**
9	27
4	12
8	24

3.

Number of 4-Leaf Clovers	Number of Leaves
1	4
3	12
5	20
4	16
9	**36**
2	**8**
8	**32**

4.

Number of Ants	Number of Legs
1	6
2	12
5	30
3	18
4	**24**
7	**42**
8	**48**

Challenge CW57

Name _____

LESSON 11.4

Fact Family Patterns

1. Fill in the missing numbers in the first three rows of the Fact Table to complete each number sentence.

Fact Table				
Blue 18 ÷ 2 = __9__	**Red** 3 × 6 = __18__	**Green** 24 ÷ 8 = __3__	**Yellow** 6 × 4 = __24__
Red 6 × __3__ = 18	**Green** 24 ÷ 3 = __8__	**Yellow** 4 × __6__ = 24	**Blue** 9 × __2__ = 18
Green 3 × __8__ = 24	**Yellow** 24 ÷ __4__ = 6	**Blue** 2 × __9__ = 18	**Red** 18 ÷ 6 = __3__
Yellow 24 ÷ 6 = 4	**Blue** 18 ÷ 9 = 2	**Red** 18 ÷ 3 = 6	**Green** 8 × 3 = 24

2. Use the colors shown below to color all the facts in the Fact Table above that belong to each fact family. **Check students' work.**

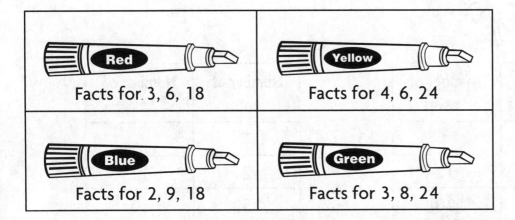

Red — Facts for 3, 6, 18
Yellow — Facts for 4, 6, 24
Blue — Facts for 2, 9, 18
Green — Facts for 3, 8, 24

3. Notice the color pattern in the Fact Table, and notice that each fact family is missing a fact. Write the missing fact from each fact family in the bottom row of the Fact Table. Arrange the facts so that the color pattern continues.

LESSON 11.5

Name _____

Picture Maker

Complete the division sentence that solves each problem. Then make a picture.

1. There are 12 kittens.
   There are 3 kittens in each basket.
   How many baskets are there in all?
   12 ÷ 3 = __4__

2. There are 15 chairs.
   There are 5 equal rows of chairs.
   How many chairs are there in each row?
   15 ÷ 5 = __3__

   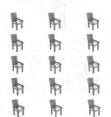

Use the picture to complete each problem. Write a division sentence for each picture.

3.	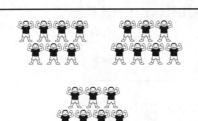	There are __21__ children. There are __3__ equal groups. How many are there in each group? __21 ÷ 3 = 7__
4.	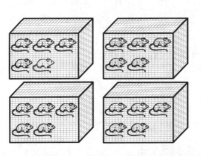	There are __20__ mice. There are __4__ cages, with the same number of mice in each cage. How many are there in each group? __20 ÷ 4 = 5__
5.	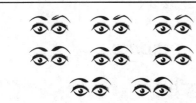	There are __16__ eyes. How many people are there? __16 ÷ 2 = 8__

Challenge CW59

Favorite Numbers

Karen, Tyler, and Daniela are friends who have made posters of their favorite numbers. Think about each of their favorite numbers and then answer the questions below.

1. Which friends have favorite numbers that can all be divided evenly by 2?

   Karen and Tyler

2. Which friends have favorite numbers that can all be divided evenly by 5?

   Daniela and Tyler

3. Which friend has favorite numbers that can all be divided evenly by both 2 and 5?

   Tyler

4. What else do Karen's favorite numbers have in common?

   Possible answer: They are all even numbers.

5. What else do Daniela's favorite numbers have in common?

   Possible answer: The ones digits are either 0 or 5.

LESSON 12.2

Name _____

The Same and Different

Divide. In each row, circle the problem that is different than the other problems in that row. Explain how the remaining problems are alike.
Answers may vary. Possible answers are given.

1. $24 \div 3 =$ __8__   2. $40 \div 4 =$ __10__   (3. $15 \div 3 =$ __5__)

   __quotients are even numbers__

(4. $36 \div 4 =$ __9__)   5. $32 \div 4 =$ __8__   6. $18 \div 3 =$ __6__

   __quotients are even numbers__

7. $20 \div 4 =$ __5__   8. $21 \div 3 =$ __7__   (9. $12 \div 3 =$ __4__)

   __quotients are odd numbers__

10. $16 \div 4 =$ __4__   11. $9 \div 3 =$ __3__   (12. $6 \div 3 =$ __2__)

   __quotient is same as divisor__

13. $30 \div 3 =$ __10__   (14. $8 \div 4 =$ __2__)   15. $20 \div 4 =$ __5__

   __quotients are divisible by 5__

16. Write two division problems that are alike in some way and one division problem that is different in some way. Have a classmate solve your problems and tell which two problems are alike and why.

    ___ ÷ ___ = ___   ___ ÷ ___ = ___   ___ ÷ ___ = ___

    Check students' work.

Challenge   CW61

Name _____

Writing Number Sentences

In each table below, the numbers in the ☐ column are dividends. The numbers in the △ column are quotients. Find the divisor that works for each table. Then, write the number sentence below the table, and complete the table. The first number sentence has been written for you.

1.

☐	△
4	2
8	4
0	**0**
6	3
10	**5**
20	**10**
14	7

Number Sentence:
☐ ÷ 2 = △

2.

☐	△
15	3
35	7
10	2
25	**5**
5	**1**
0	**0**
45	9

Number Sentence:
☐ ÷ 5 = △

3.

☐	△
20	5
16	4
4	1
12	**3**
8	2
36	9
0	0

Number Sentence:
☐ ÷ 4 = △

4.

☐	△
12	2
24	4
6	1
30	5
42	7
0	**0**
18	3

Number Sentence:
☐ ÷ 6 = △

5.

☐	△
12	4
27	9
6	2
3	1
0	0
18	**6**
9	3

Number Sentence:
☐ ÷ 3 = △

6.

☐	△
4	4
8	8
7	7
5	**5**
0	**0**
3	3
2	**2**

Number Sentence:
☐ ÷ 1 = △

CW62 Challenge

LESSON 12.4

Name _____

Write a Problem

Write an expression that describes each picture. Then write a word problem to go with the picture and the expression.
Answers may vary. Possible answers are given.

1.

    Expression: _____ 8 ÷ 2 _____

    The Problem

    A pet store has 8 kittens. It keeps an equal number of kittens in each of two cages. How many kittens are in each cage?

2.

    Expression: _____ 6 + 5 _____

    The Problem

    Six birds are on the ground. Five more birds join them. How many birds are on the ground now?

3.

    Expression: _____ 9 − 3 _____

    The Problem

    The boy had 9 balloons. Three of the balloons blew away. How many balloons does he have now?

4.

    Expression: _____ 5 × 8 _____

    The Problem

    The grocery store displays apples in 5 equal rows of 8 apples. How many apples are in the display?

Challenge CW63

Name _____

LESSON 12.5

Solving Problems at the Aquarium

Aquarium	Admission	Sea Lion Show
	$6 adults $4 children under 12 $45 Family membership— free admission for one year	10:30, 12:00, 1:30, 3:00

1. Mr. and Mrs. Young and their 6-year-old triplets go to the aquarium. How much do they pay?

   $24.00 (or $45 for 1 yr. membership)

2. The sea lion show lasts 45 minutes. How much time is there between shows?

   45 min

3. The theater where the sea lions perform can seat 600 people. There are 475 people sitting in the theater for the 12:00 show. How many more people can be seated before the theater is full?

   125 more people

4. Mr. Ruiz buys a family membership. He goes to the aquarium with his 4-year-old son 6 times during the year. How much money does he save?

   $15.00

5. A class of 24 students visits the aquarium. They divide into 4 groups. How many students are in each group?

   6 students

6. John buys a book about sharks for $4.95 and a shell for $1.35. How much money does he spend?

   $6.30

7. Meg counts 12 starfish and 9 hermit crabs in a display. How many more starfish are there than hermit crabs?

   3 more starfish

8. Jesse learned that a seahorse egg hatches in 50 to 60 days. About how many weeks is this?

   about 8 weeks

CW64 Challenge

Name _____

 LESSON 13.1

Fun with Facts

Follow the arrows to solve each problem. Write the answer inside the empty box. You may use a multiplication table to help you multiply and divide.

1. [36] → [÷6] → [×4] → [÷8] → [3]

2. [48] → [÷8] → [+15] → [÷7] → [3]

3. [63] → [÷7] → [×2] → [÷6] → [3]

4. [56] → [÷7] → [−2] → [÷6] → [1]

5. [35] → [÷7] → [+27] → [÷8] → [4]

6. [16] → [÷8] → [×6] → [−4] → [8]

7. [8] → [×3] → [÷6] → [+4] → [8]

Write an operation and a number in the empty box.

8. [72] → [÷8] → [÷3] → [×7] → [21]
                                                         or + 18

9. [24] → [÷6] → [×9] → [÷6] → [6]
                                                         or − 30

10. [8] → [×8] → [−8] → [÷7] → [8]
                                                        or − 48

Make up your own problems. Use at least one multiplication or division step in each. **Check students' work. Answers will vary.**

11. [ ] → [ ] → [ ] → [ ] → [ ]

12. [ ] → [ ] → [ ] → [ ] → [ ]

Challenge CW65

Name _____

Dot-to-Dot Division

Divide. Connect the dots in order from *least* to *greatest* quotient.

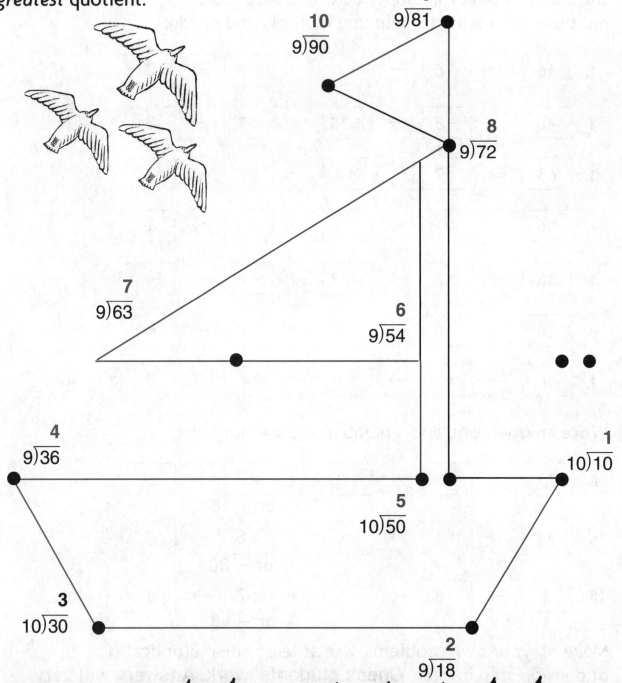

On grid paper, design your own connect-the-dots picture using the division facts you know. Have a friend solve.

CW66   Challenge

The Quotient Game

Play with a partner.

Materials:
2 game tokens
2 sets of number cards with the numbers 1–10

How to Play:
Place the game tokens on **Start**. Shuffle the number cards and turn them face down.

Take turns drawing a number card. Move forward to the closest space with a quotient that matches the number on the card. The first player to reach **Finish** wins!

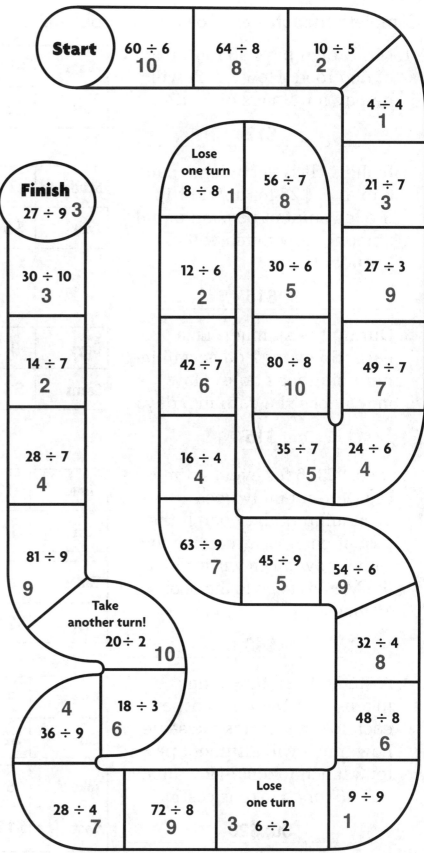

Name _____

> LESSON 13.4

What's the Cost?

Complete the table and solve each problem.

1. Mr. Brown pays $3 for 1 bag of dog food. How much will he pay for 4 bags of food?

   $12

Bags	1	2	3	4	5
Cost	$3	$6	$9	$12	$15

2. In Jill's class, 5 students paid a total of $25 for a field trip. If each student pays an equal amount, how much will 3 students pay?

   $15

Students	1	2	3	4	5
Cost	$5	$10	$15	$20	$25

3. During the summer, Liza earns $6 every 2 days walking her neighbor's dogs. How much does she earn in 5 days?

   $15

Days	1	2	3	4	5
Earns	$3	$6	$9	$12	$15

4. It costs $16 for every 4 times the Moran family goes swimming at the pool. If the cost of each visit remains the same, how much will it cost for the Morans to go to the pool 10 times?

   $40

Times	1	2	3	4	5
Cost	$4	$8	$12	$16	$20

Times	6	7	8	9	10
Cost	$24	$28	$32	$36	$40

5. A student can buy 9 lunch tokens for $18. If the cost of each token remains the same, how much will a student pay for 3 lunch tokens? How much will 10 lunch tokens cost?

   $6; $20

Tokens	1	2	3	4	5
Cost	$2	$4	$6	$8	$10

Tokens	6	7	8	9	10
Cost	$12	$14	$16	$18	$20

CW68 Challenge

What Number Am I?

Read each number riddle. Use the clues to identify each number.

1. If you add 7 to me and then divide the result by me, the answer is 8. What number am I?

   _____1_____

2. Multiply me by 5 and you get a number that is 5 more than 20. What number am I?

   _____5_____

3. Divide me by 3, or multiply me by 6. The answer will be the same. What number am I?

   _____0_____

4. When you multiply me by myself, you get me again! What number am I?

   _____1 or 0_____

5. Divide 12 by 3 and you get me. Divide 12 by me and you get 3. What number am I?

   _____4_____

6. If you multiply any number by me, the sum of the digits in the product equals me. What number am I?

   _____9_____

7. If you divide 30 by me, the answer is 3 doubled. What number am I?

   _____5_____

8. If you divide any number by me, your answer will be that number again! What number am I?

   _____1_____

9. Write your own number riddle. After solving it yourself, ask a classmate to try it.

   Answers will vary.

Sara's System

Sara collects dolls from around the world. The table shows what kinds of dolls she has in her collection.

SARA'S DOLLS			
Number	Country	Gender	Adult or Child
2	Australia	Male	Adult
3	Australia	Female	Adult
5	Australia	Female	Child
1	United States	Female	Adult
4	United States	Male	Child
5	United States	Female	Child
1	Sweden	Male	Adult
2	Japan	Male	Adult
3	Japan	Female	Adult
4	Japan	Male	Child
5	Mexico	Female	Child
3	Israel	Male	Adult
2	Israel	Female	Child

Sara sorted her dolls into four boxes. All the dolls in each box have two things in common. Write which type of doll is in each box.

Male Adults = 8; Female Adults = 7; Male Children = 8;

Female Children = 17

Name _____

LESSON 14.2

You Decide

You work at a grocery store.

You must decide what brand of crackers to order.

A survey is conducted to help you.

Think about the price of the crackers and the number of votes when you decide.

CRACKERS		
Brand of Crackers	Number of Votes	Price per Bag
Wavy	23	30¢
Light 'n' Salty	13	36¢
Crispy Crunchies	45	27¢
Toasties	22	20¢
Frickles	61	23¢
Ring-a-Ling	48	18¢
Goodies	27	25¢

Examine the table. Pick three brands to order.
Explain your decision.

<u>**Possible answer: I would order Frickles, Crispy Crunchies, and**</u>

<u>**Goodies because they are priced in the middle and had**</u>

<u>**a large number of votes.**</u>

Challenge CW71

Name _____

LESSON 14.3

What's the Question?

These answers need questions! Use the information and the data in the tables to write a question for each answer.

For 1–2, use the tally table. **Answers may vary. Check students' work.**

1. Answer: birds and squirrels

   Question: _____
   _____
   _____

2. Answer: 14 animals

   Question: _____
   _____
   _____

Sarah likes to watch animals in the park. The tally table shows the numbers of animals she saw yesterday.

Animals in the Park	
Animal	Tally
bird	𝍢
rabbit	///
squirrel	𝍢
snake	/

For 3–4, use the frequency table.

3. Answer: Tuesday

   Question: _____
   _____
   _____

4. Answer: 15 minutes

   Question: _____
   _____
   _____

Trevor runs every day of the week. The frequency table shows the number of minutes he ran each day.

Minutes of Running	
Day	Minutes
Monday	30
Tuesday	45
Wednesday	20
Thursday	35
Friday	30

CW72   **Challenge**

Month to Month

Use the bar graphs. Write the letters A, B, or C to answer each question.

1. Which graph shows that Luis read 6 books in one month?
   B

2. Which graph shows the month that Sue read the least number of books?
   C

3. Which graphs show the two months that Tia read the same number of books?
   A and B

4. Which graph shows that the students read a total number of 17 books?
   C

5. Which graph shows the month that Sue read 2 more books than Jack?
   B

6. Which graph shows the greatest number of books read in one month?
   A

7. Which graph shows the month that Tia read the most books?
   C

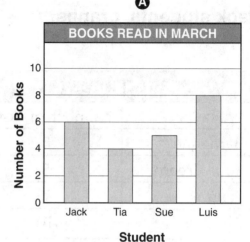
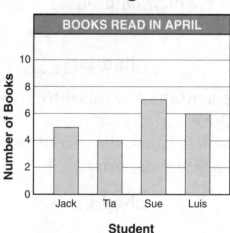
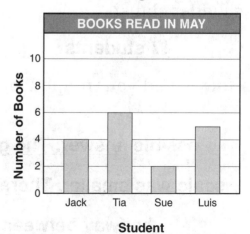

Challenge CW73

Name _____

LESSON 14.5

Barbeque or Cheddar?

Use the data in the table to make a bar graph.
Use your bar graph to answer the questions.
Check students' graphs.

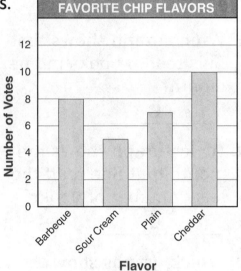

FAVORITE CHIP FLAVORS	
Flavor	Number of Votes
Barbeque	8
Sour Cream	5
Plain	7
Cheddar	10

1. For which chip flavor is the bar the longest?

   _____**cheddar**_____

2. For which chip flavor is the bar the shortest?

   _____**sour cream**_____

3. If a student wanted to bring chips for a class party, which flavor do you think they should bring?

   _____**cheddar**_____

4. How many more students like barbeque chips than sour cream chips?

   _____**3 students**_____

5. How many students like either plain chips or cheddar chips?

   _____**17 students**_____

6. How many students in all were surveyed?

   _____**30 students**_____

7. How would your graph look different if you had used a scale of 1?

   Possible answer: The graph might have been taller if the

   scale was smaller. There wouldn't be any bars that stopped

   halfway between two numbers on the scale.

CW74 Challenge

Name _____

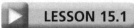

No Rulers Allowed!

Measure and cut a 5-inch strip and a 2-inch strip from a piece of paper.

5 inches

2 inches

Use the two strips to measure the drawings in 1–4 to the nearest half inch. Do **not** use a ruler.

Hints:

- If you fold each strip in half, you will have two more measuring strips. Half of the 5-inch strip will be $2\frac{1}{2}$ inches long. Half of the 2-inch strip will be ____**1 inch long**____.

- You can compare or combine the rulers to form different lengths.

1.    **3 in.**

2.   **$1\frac{1}{2}$ in.**

3.   **1 in.**

4.    **$4\frac{1}{2}$ in.**

5. Tell how you measured the pen in Problem 1.

   Possible answer: I used the 2-inch strip. Then I folded the strip in half to make a 1-inch strip. The pen measured 3 inches.

Challenge CW75

Name _____

LESSON 15.2

Choose the Best Unit

1. What is the best unit for measuring each item on the quilt? Use the key to color each triangle on the quilt.
Check students' work.

Key	
inch	red
foot or yard	yellow
mile	green

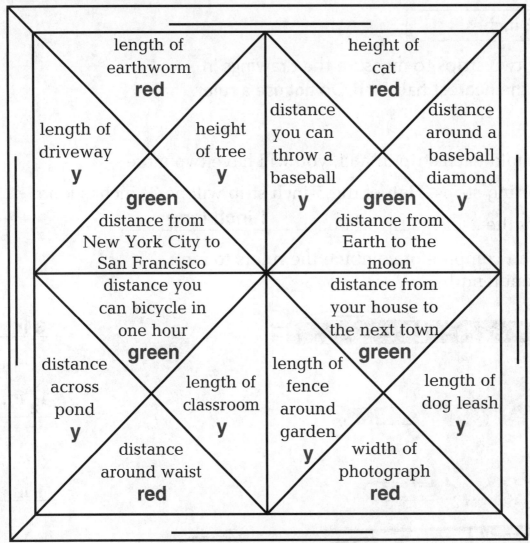

2. In the top border of the quilt, write the name of an item to be measured in feet or yards. Color the top border yellow.
Answers will vary. Check students' answers.
3. In the bottom border of the quilt, write the name of an item to be measured in miles. Color the bottom border green. **Answers will vary. Check students' answers.**
4. In each side border of the quilt, write the name of an item to be measured in inches. Color the side borders red. **Answers will vary. Check students' answers.**

CW76 Challenge

Name _____

LESSON 15.3

Graphing Length

Five students wanted to see how far they could jump from standing still. Each of their best jumps is shown in the table.

Student	Jump
Alice	1 foot 2 inches
Bob	1 foot 8 inches
Donna	17 inches
Elia	1 foot 4 inches
Juan	13 inches

JUMPS OF STUDENTS

(Inches axis: 0 to 40; Students axis)

1. Use the table to make a bar graph.
   Check students' graphs.
2. What is the difference between the least and greatest jumps shown in the graph? __7 inches__
3. How many inches did the students jump altogether? __80 inches__
4. Order the students from their shortest jump to their longest jump.

   Juan, Alice, Elia, Donna, Bob

Challenge CW77

Name _____

> LESSON 15.4

Centimeter Estimation Game

Play with a partner.

Materials:

- table shown below for each player
- centimeter ruler

How to Play:

Step 1 Work with your partner to identify 10 objects or distances in the classroom that you will measure in centimeters. Record them in the first column of your tables.

Step 2 Work by yourself to estimate the length of each object or distance in centimeters. Record your estimates in the second column of your table.

Step 3 Work with your partner to measure the length of each object or distance to the nearest centimeter. Record these measurements in the third column of your tables.

Step 4 Work by yourself to find the difference between the estimated length and actual length of each object or distance. Record these differences in the fourth column of your table.

Step 5 Work by yourself to find the sum of the differences in the fourth column. The player with the lower sum wins!

Object/Distance	Estimated Length	Actual Length	Difference

CW78 Challenge

LESSON 15.5

Name _____

What's the Order?

Order the lengths from *least* to *greatest*.

1. 305 cm, 31 km, 3 m, 35 mm

   35 mm, 3 m, 305 cm, 31 km

2. 295 cm, 2 m, 2 km, 20 mm

   20 mm, 2 m, 295 cm, 2 km

3. 15 km, 15 m, 5 cm, 51 mm

   5 cm, 51 mm, 15 m, 15 km

4. 8 m, 878 cm, 78 m, 87 mm

   87 mm, 8 m, 878 cm, 78 m

5. 355 cm, 5 m, 535 cm, 35 m

   355 cm, 5 m, 535 cm, 35 m

6. 986 cm, 98 km, 86 m, 89 cm, 9 m

   89 cm, 9 m, 986 cm, 86 m, 98 km

7. 92 km, 29 m, 920 cm, 290 cm, 229 mm

   229 mm, 290 cm, 92 dm, 920 cm, 29 m, 92 km

8. Think of two times when metric units of length might be used.

   Possible answer: in a science

   class; in sports

9. Write a problem using the measurements listed in one of the problems above. Solve.

   Check students' problems and solutions.

Challenge CW79

Name _____

LESSON 15.6

Take a Hike

There are 5 teams of campers: Red, Yellow, Blue, Green, and Orange. They are hiking from Camp Evergreen to Pinewood Camp along a trail. The trail is 5 kilometers long and each team is at a different stop along the trail.

Use logical reasoning to find out which team is at which trail stop.

- The Red Team has hiked 3 kilometers.
- The Blue Team is 2 kilometers ahead of the Red Team.
- The Green Team has hiked 2 kilometers less than the Red Team.
- The Orange Team is 1 kilometer ahead of the Red Team.

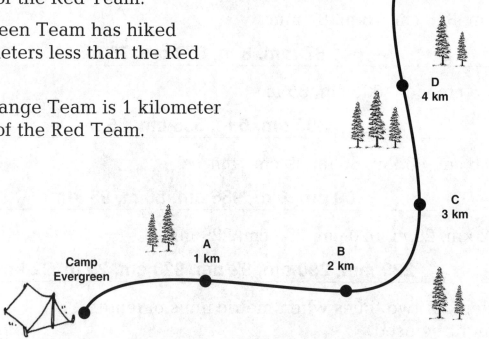

1. Which team is at trail stop D? __Orange__

2. Write the names of the teams in order from the least kilometers hiked to the most kilometers hiked.

   Green, Yellow, Red, Orange, Blue

3. How many kilometers have the teams hiked in all? Write a number sentence to explain your answer.

   15 kilometers; 5 + 4 + 3 + 2 + 1 = 15

CW80 Challenge

What's the Angle?

Play with a partner.

Materials:

- 12 index cards, ruler

How to Play:

- Draw angles on the back of each index card. Be sure to have 4 that are *right angles*, 4 that are *acute angles*, and 4 that are *obtuse angles*.

- Shuffle the cards and lay them, face down, in 3 rows with 4 cards in each row.

- The object is to win cards by matching angles.

- The first player turns 2 cards over. It they are both the same type of angle, the player wins the cards. If they are not the same type, turn them back over and the next player takes a turn.

- The player with the most cards at the end wins.

Examples:

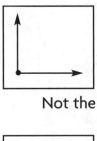 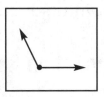
Not the same type

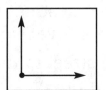 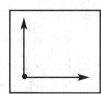
Same type

Not the same type

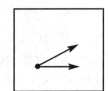
Same type

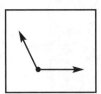 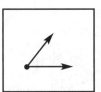
Not the same type

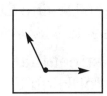
Same type

Challenge

Name _____

LESSON 16.2

Mapmaker, Mapmaker, Make Me a Map!

Use the directions below and what you know about lines and angles to label the streets on the map.

Check students' answers.

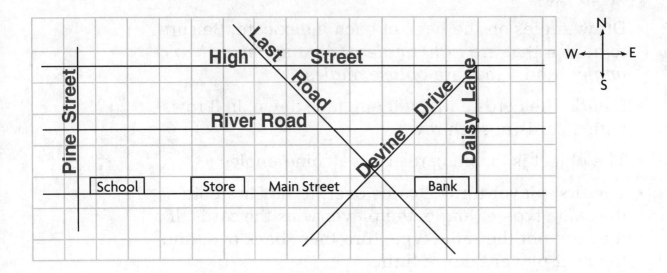

1. Label the first street to the north of and not intersecting Main Street *River Road*.
2. Label the street to the north of and not intersecting River Road *High Street*.
3. Label the street to the east of the bank and makes a right angle to Main Street *Daisy Lane*.
4. Label the street to the west of the school *Pine Street*. It forms a right angle with Main Street and intersects High Street.
5. Label the street that meets Daisy Lane and High Street *Devine Drive*. It runs southwest and intersects Main Street east of the store.
6. Label the street that forms an obtuse angle on the south side of High Street *Last Road*. It intersects Main Street west of the bank.

Name _____

LESSON 16.3

Polygon Puzzle

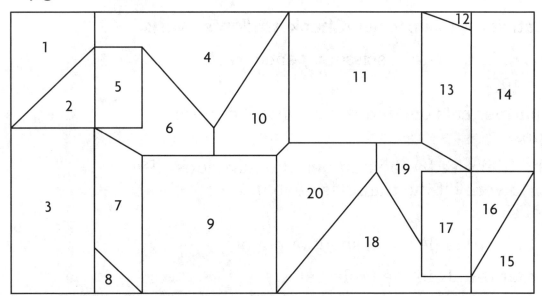

Answer the following questions and then follow the directions.

1. Write the numbers of the polygons that have 4 sides and 4 angles.

   1, 3, 5, 7, 9, 11, 13, 14, 15, 17

2. Write the numbers of the polygons that have 3 sides and 3 angles.

   2, 8, 12, 16

3. Write the numbers of the polygons that have 5 or more sides and 5 or more angles.

   4, 6, 10, 18, 19, 20

4. Use the chart to color the polygons.

5. Make your own design using polygons. Color the design. **Check students' designs.**

Numbers	Color
1, 15	Black
2, 16	Red
3, 14	Yellow
7, 13	Green
8, 12	Orange
5, 17	Purple
6, 19	Brown
10, 20	Blue
9, 11	Pink
4, 18	Light Blue

Challenge CW83

Name _____

LESSON 16.4

Triangle Tally

Do this activity with a partner. **Check students' work.**

Materials: 3 index cards, scissors, pencil, ruler, sheet of paper

- Each partner cuts one index card into 3 different triangles.
- Sort the triangles by the number of equal sides. Use a ruler to verify if the triangles have 0, 2, or 3 equal sides.
- Trace the 6 triangles on a sheet of paper.
- Tally your results in the table below.

SIDES OF THE TRIANGLES

0 equal sides	2 equal sides	3 equal sides

- Sort the triangles by their angles. Use the corner of the third index card to see if each triangle has one right angle, one obtuse angle, or three acute angles.
- Tally your results in the table below.

ANGLES OF THE TRIANGLES

1 right angle	1 obtuse angle	3 acute angles

CW84 Challenge

Quadrilateral Puzzles

Read the clues. Color the figures. Write the name of each figure.

1. If a quadrilateral has 1 pair of sides the same distance apart and 2 right angles, color it red.

2. If a quadrilateral has 4 right angles and 2 pairs of equal sides, color it blue.

3. If a quadrilateral has no equal sides and no right angles, color it green.

4. If a quadrilateral has 4 right angles and 4 equal sides, color it purple.

5. If a quadrilateral has one pair of sides the same distance apart and no right angles, color it brown.

6. If a quadrilateral has 6 right angles, color it black. **There should be no quadrilaterals colored black.**

7. If a quadrilateral has 2 pairs of sides the same distance apart, no right angles, and 2 pairs of equal sides, color it pink.

8. If a quadrilateral has 1 right angle and 1 pair of parallel sides, color it gray. **There should be no quadrilaterals colored gray.**

9. If a quadrilateral has 4 right angles and no equal sides, color it yellow. **There should be no quadrilaterals colored yellow.**

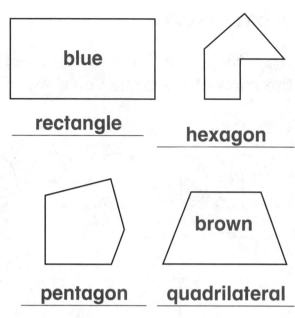

blue — rectangle   hexagon

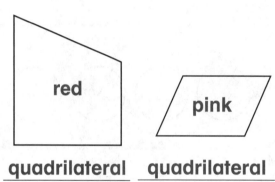

pentagon   brown — quadrilateral

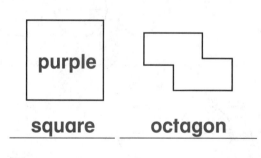

red — quadrilateral   pink — quadrilateral

purple — square   octagon

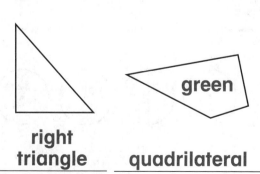

right triangle   green — quadrilateral

Challenge CW85

Name _____

LESSON 16.6

What's Next?

Draw the circle that will come next in the pattern. Name the parts of the circle you draw.

1.

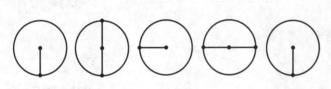 | center, diameter

2.

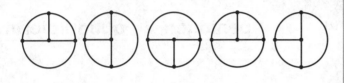 | center, diameter, radius

3.

 | center

4.

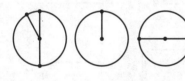 | center, diameter

5.

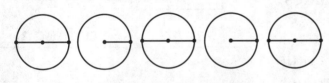 | center, radius

6.

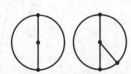 | center, radius, diameter

CW86 Challenge

Name _____

LESSON 16.7

Missing Labels

Label the sets in each Venn diagram below.

1. **Figures That Have Straight Sides** / **Figures That Have Curved Sides**

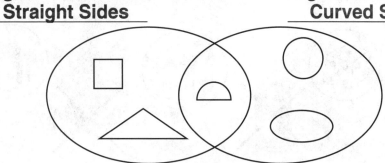

2. **Multiples of 3** / **Multiples of 4**

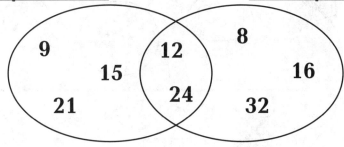

3. **Intersecting Lines** / **Lines That Don't Intersect**

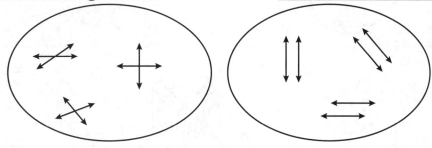

4. Draw your own Venn diagram, but do not label the sets. You may use geometric figures or numbers. Ask a classmate to label the sets.
   Check students' diagrams.

Challenge CW87

Name _____

Solid Match

Circle the objects that look like the solid figure.
Put an X on the figures that do not.

1.

2.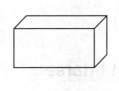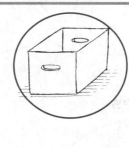

3.

4.

5.

LESSON 17.1

CW88 Challenge

Name _____

LESSON 17.2

The Missing Half

Each figure below was once a cube, rectangular prism, sphere, cone, or cylinder. Each has been cut in half with one of the halves removed.

- Complete the solid figure.
- Name the solid figure. **Check students' drawings.**

1.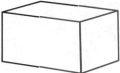

   rectangular prism

2.

   sphere

3.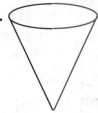

   cone

4.

   cylinder

5.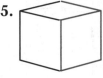

   cube

6.

   cone

Each figure below is made from two solids. One solid has been cut in half with one of the halves removed. **Check students' drawings.**

- Complete the figure. Name the solid figures.

7.

   sphere; cube

8.

   cone, cylinder

Challenge   CW89

Name _____

LESSON 17.3

Folding Solid Figures

You can make solid figures with paper. Look at the patterns on this page.

- Copy the patterns onto grid paper.
- Cut on the solid lines.
- Fold on the dotted lines.
- Use tape to hold your solid figure together.

What figures did you make?

_____a rectangular prism and a cube_____

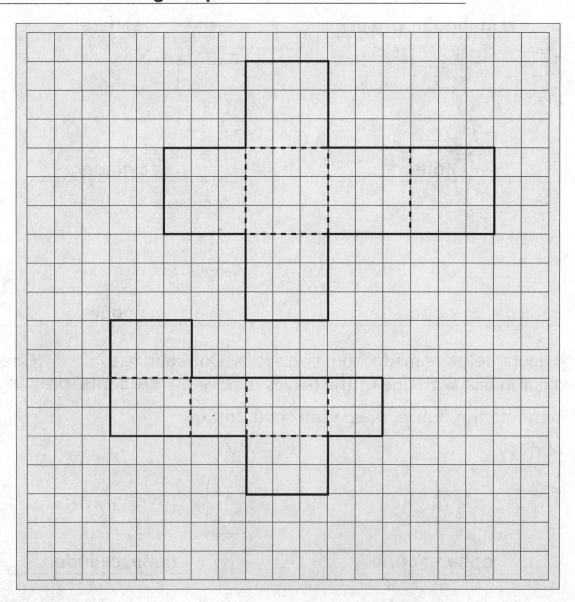

CW90 Challenge

Name _____

LESSON 17.4

Different Views

You can put cubes together in many different ways to make new solid figures. When you do this, the top, front, and side views of the new figure may be different.

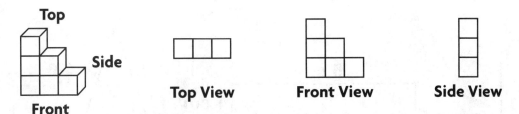

Sketch the top, bottom, and side views of each figure. If you need help, build the figure out of cubes and look at it from different views.

	Top View	Front View	Side View
1.			
2.			
3.			
4.			

Challenge CW91

Name _____

▶ LESSON 18.1

Find the Perimeter

Estimate the perimeter of each figure in centimeters (cm). Then use a centimeter ruler to find the actual perimeter. **Estimates will vary.**

⊢⊣ 1 cm

1.
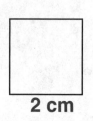

estimate: _____ cm

perimeter: __8__ cm

2.
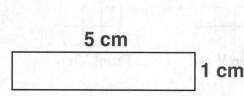

estimate: _____ cm

perimeter: __12__ cm

3.
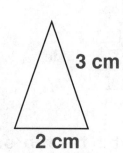

estimate: _____ cm

perimeter: __8__ cm

4.
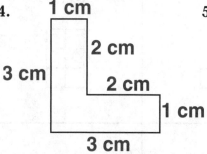

estimate: _____ cm

perimeter: __12__ cm

5.
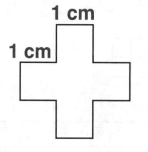

estimate: _____ cm

perimeter: __12__ cm

6.
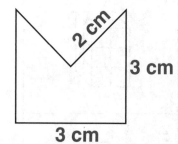

estimate: _____ cm

perimeter: __13__ cm

7.
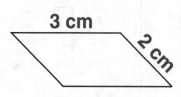

estimate: _____ cm

perimeter: __10__ cm

8.
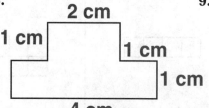

estimate: _____ cm

perimeter: __12__ cm

9.
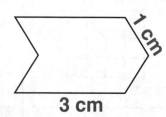

estimate: _____ cm

perimeter: __10__ cm

CW92 **Challenge**

Name _____

LESSON 18.2

What's Missing?

Use a formula to solve each problem. Find the missing length, width, or side.

1.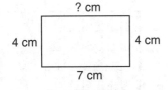

   $P = l + w + l + w$
   $22 = 7 + 4 + \underline{\ 7\ } + 4$

2.

   $P = l + w + l + w$
   $22 = 9 + \underline{\ 2\ } + 9 + \underline{\ 2\ }$

3.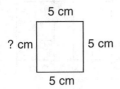

   $P = s + s + s + s$
   $20 = 5 + \underline{\ 5\ } + 5 + 5$

4.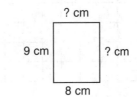

   $P = l + w + l + w$
   $34 = 9 + \underline{\ 8\ } + \underline{\ 9\ } + 8$

5.

   $P = s + s + s + s$
   $4 = \underline{\ 1\ } + \underline{\ 1\ } + \underline{\ 1\ } + \underline{\ 1\ }$

6.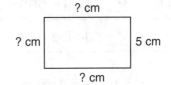

   $P = l + w + l + w$
   $24 = \underline{\ 7\ } + \underline{\ 5\ } + \underline{\ 7\ } + 5$

Challenge CW93

Name _____

▶ LESSON 18.3

Areas in Town

Find the area of each building in the town drawn on the grid below. Record your findings in the table.

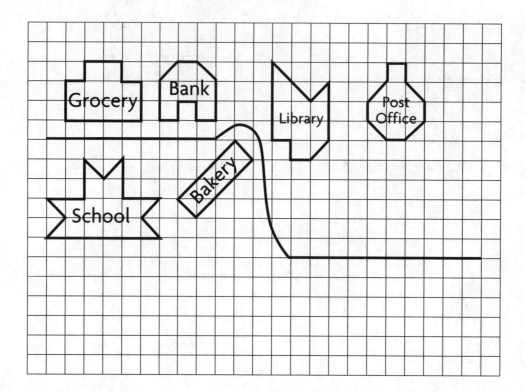

	Building	Area
1.	Grocery	__10__ square units
2.	Bank	__7__ square units
3.	School	__13__ square units
4.	Library	__10__ square units
5.	Post Office	__8__ square units
6.	Bakery	__6__ square units

7. Add a toy shop to the town. Make it have an area of 9 square units. **Drawings will vary. Check students' drawings.**

8. Add a restaurant to the town. Make it have the same area as the toy shop but a different shape. **Drawings will vary. Check students' drawings.**

Name _____

LESSON 18.4

What's My Measure?

Solve each problem by finding the length, width, or area of a rectangle. Draw a diagram of each rectangle. Label the sides with inches or feet. Check students' drawings.

1. Jason has a rug that is 2 feet wide. The length of the rug is 3 times the size of the width. What is the area of Jason's rug?

   12 square feet

   6 ft
   2 ft

2. Katie needs to fill a space in her scrapbook with a picture. The space can fit a picture that is 3 inches tall and has an area of 12 square inches. What will the length of the picture be?

   4 inches

   4 in.
   3 in.

3. Cimona is covering the lid of a box with paper. The lid has an area of 49 square inches. She needs to cut her paper 2 inches larger so it will wrap around the edges of the lid. What will be the area of the paper?

   81 square inches

   9 in.
   9 in.

4. Mr. Parker put fresh soil in his garden. He covered 50 square feet of the garden, but needs 4 more square feet of soil to finish the job. The length of his garden is 9 feet. What is the width of the garden?

   6 feet

   9 ft
   6 ft

Challenge CW95

Painting Project

Amanda and her father are painting the walls of a playroom. One of the walls has a window. Another wall has a door, which they are not painting.

Find the area that needs to be painted on each of the four walls. All measurements are given in feet.

1.

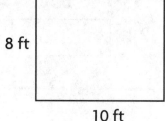

area = __80__ square feet

2.

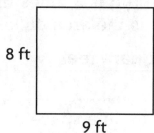

area = __72__ square feet

3.

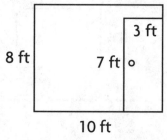

area = __59__ square feet

4.

area = __60__ square feet

5. What is the total area that needs to be painted?

   __271 sq ft__

6. The label on the can of paint says that 1 quart covers about 100 square feet. How many quarts of paint will Amanda and her father need to paint the walls with 2 coats of paint?

   __6 qt__

7. Amanda wants to put a wallpaper border around the room near the ceiling. How many feet of wallpaper border does she need to go all the way around the room?

   __38 ft__

Name _____

LESSON 19.1

Continue the Pattern

Find a pattern unit on each grid.
Use it to continue each pattern.

1.
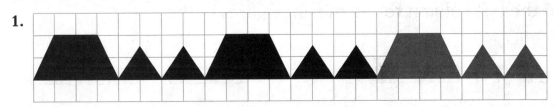

2.
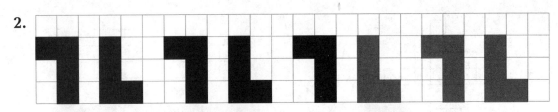

3.

4.
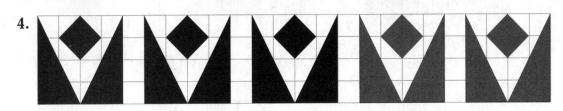

5.
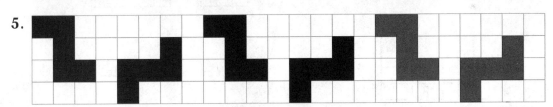

6.
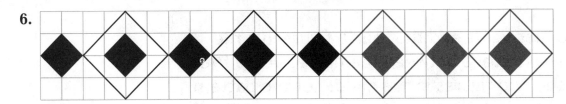

7. Use the grid to make your own pattern. **Check students' work.**

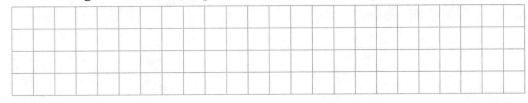

Challenge CW97

Missing Tiles

Two figures are missing from the middle of each tile pattern in the left column. Draw a line from each pattern in the left column to the missing figures in the right column.

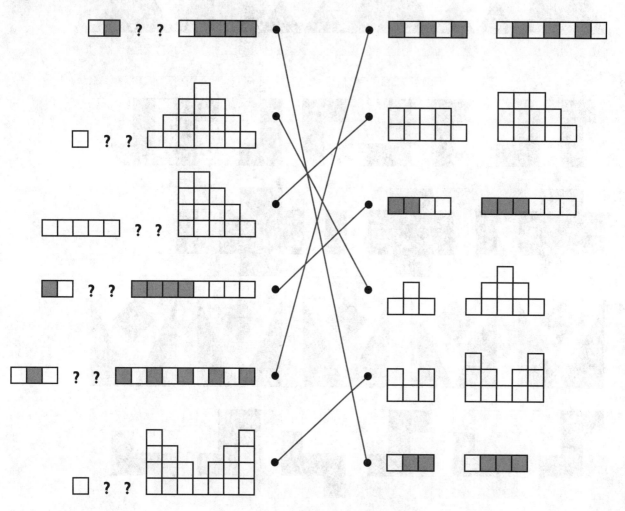

Draw the first figure and the fourth figure in a tile pattern. Leave space between the figures. Ask a friend to draw the two figures in the middle of the pattern.

CW98 Challenge

Traveling Birds

Riddle: Why do birds fly south?

Find the missing numbers.

1. 93, 87, 81, 75, __69__, 63, __57__, __51__
                             T   A   S

2. 30, 37, 44, 51, __58__, __65__, 72, __79__
                           E   O   H

3. 245, 257, 269, 281, __293__, 305, __317__, __329__
                                   Y   A   L

4. 635, 626, 617, 608, __599__, __590__, 581, __572__
                                   T   K   T

5. 680, 655, 630, 605, __580__, __555__, 530, __505__
                                   B   E   K

6. 1,326; 1,426; 1,526; __1,626__; 1,726; __1,826__; __1,926__
                                       N   U   I

7. 2,036; 2,016; 1,996; 1,976; __1,956__; __1,936__; 1,916; __1,896__; __1,876__
                                           O   T   D   E

Now solve the riddle. Match each letter to the numbers you found in the number patterns. Write the letters on the lines below.

T	H	E	Y		D	O	N	'	T
599	79	555	293		1,896	65	1,626		572

L	I	K	E		T	O
329	1,926	590	1,876		69	1,956

T	A	K	E		A		B	U	S
1,936	57	505	58		317		580	1,826	51

Name _____

LESSON 19.4

Missing Numbers

Complete the patterns below. Then find the answers in the number search. Circle the answers. The numbers can be found going up, across, down, backward, and diagonally.

1. 225, 255, 285, __315__, 345, __375__, __405__

2. 5,423; 5,413; 5,403; __5,393__; __5,383__; 5,373; __5,363__

3. 764, __744__, __724__, 704, 684, 664, __644__

4. 3,240; __3,255__; __3,270__; __3,285__; 3,300; 3,315; 3,330

5. 160, 184, 208, __232__, __256__, 280, __304__

6. 948, __916__, __884__, __852__, 820, 788, 756

7. 3,460; 3,435; 3,410; __3,385__; 3,360; __3,335__; __3,310__

8. 2,154; 2,187; 2,220; __2,253__; __2,286__; 2,319; __2,352__

3	2	5	3	6	3	6	5	1	3
7	2	6	0	9	1	4	3	2	3
4	8	7	1	9	3	2	2	1	3
4	6	9	0	8	5	7	7	3	5
0	0	1	3	3	3	8	5	4	2
8	9	5	2	0	2	1	3	6	7
0	4	2	5	6	8	2	9	1	5
1	7	9	2	6	5	5	3	5	3
3	7	5	4	0	3	8	2	0	4
3	1	4	8	8	4	3	5	2	2

CW100 Challenge

Fetching Fractions

Find the word name in Column 2 for each fraction in Column 1. Then write the word name's circled letter on the line in front of the fraction.

Column 1

__L__ 1. $\frac{1}{2}$
__S__ 2. $\frac{3}{4}$
__A__ 3. $\frac{2}{5}$
__W__ 4. $\frac{1}{3}$
__O__ 5. $\frac{6}{10}$
__C__ 6. $\frac{2}{9}$
__F__ 7. $\frac{1}{4}$
__E__ 8. $\frac{5}{9}$
__N__ 9. $\frac{4}{8}$
__H__ 10. $\frac{1}{10}$
__T__ 11. $\frac{5}{6}$
__P__ 12. $\frac{1}{5}$
__I__ 13. $\frac{7}{8}$
__R__ 14. $\frac{8}{9}$

Column 2

(A) two fifths
(C) two ninths
(E) five divided by nine
(F) one out of four
(H) one tenth
(I) seven divided by eight
(L) one out of two
(N) four eighths
(O) six tenths
(P) one out of five
(R) eight ninths
(S) three divided by four
(T) five sixths
(W) one out of three

Now decode the sentence below. The numbers tell you where to look in Column 1. Write the letter that is on the blank in front of the number.

$\frac{A}{3}$ $\frac{F}{7}$ $\frac{R}{14}$ $\frac{A}{3}$ $\frac{C}{6}$ $\frac{T}{11}$ $\frac{I}{13}$ $\frac{O}{5}$ $\frac{N}{9}$ $\frac{I}{13}$ $\frac{S}{2}$

$\frac{P}{12}$ $\frac{A}{3}$ $\frac{R}{14}$ $\frac{T}{11}$ $\frac{O}{5}$ $\frac{F}{7}$ $\frac{A}{3}$ $\frac{W}{4}$ $\frac{H}{10}$ $\frac{O}{5}$ $\frac{L}{1}$ $\frac{E}{8}$.

Challenge CW101

Name _____

LESSON 20.2

Color the Apples

How could you color $\frac{3}{4}$ of 8 apples red?

Look at the *denominator*.

It tells you to make 4 equal parts.

Divide the 8 apples into 4 equal groups.

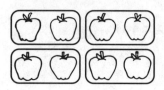

Each group has 2 apples.

Look at the *numerator*.

It tells you to color 3 of the groups.

Color 3 of the 4 groups.

The picture shows 3 groups or 6 apples shaded.

So, $\frac{3}{4}$ of 8 = 6.

Color the apples to show the part of the group the fraction names. Solve. **Check students' coloring.**

1.

   Color $\frac{1}{4}$ red.

   $\frac{1}{4}$ of 12 = ___3___

2.

   Color $\frac{3}{4}$ green.

   $\frac{3}{4}$ of 12 = ___9___

3.

   Color $\frac{1}{5}$ green.

   $\frac{1}{5}$ of 15 = ___3___

4.

   Color $\frac{2}{5}$ red.

   $\frac{2}{5}$ of 15 = ___6___

CW102 Challenge

Name _____

LESSON 20.3

Fraction Squares

Fill in the empty squares to have correct addition across and down.

1.

$\frac{1}{4}$	+	$\frac{1}{4}$	=	$\frac{2}{4}$
+		+		+
$\frac{1}{4}$	+	$\frac{1}{4}$	=	$\frac{2}{4}$
=		=		=
$\frac{2}{4}$	+	$\frac{2}{4}$	=	$\frac{4}{4}$

2.

$\frac{1}{6}$	+	$\frac{2}{6}$	=	$\frac{3}{6}$
+		+		+
$\frac{1}{6}$	+	$\frac{1}{6}$	=	$\frac{2}{6}$
=		=		=
$\frac{2}{6}$	+	$\frac{3}{6}$	=	$\frac{5}{6}$

3.

$\frac{2}{10}$	+	$\frac{4}{10}$	=	$\frac{6}{10}$
+		+		+
$\frac{2}{10}$	+	$\frac{1}{10}$	=	$\frac{3}{10}$
=		=		=
$\frac{4}{10}$	+	$\frac{5}{10}$	=	$\frac{9}{10}$

4.

$\frac{4}{12}$	+	$\frac{3}{12}$	=	$\frac{7}{12}$
+		+		+
$\frac{3}{12}$	+	$\frac{1}{12}$	=	$\frac{4}{12}$
=		=		=
$\frac{7}{12}$	+	$\frac{4}{12}$	=	$\frac{11}{12}$

5.

$\frac{6}{20}$	+	$\frac{5}{20}$	=	$\frac{11}{20}$
+		+		+
$\frac{4}{20}$	+	$\frac{3}{20}$	=	$\frac{7}{20}$
=		=		=
$\frac{10}{20}$	+	$\frac{8}{20}$	=	$\frac{18}{20}$

6.

$\frac{2}{14}$	+	$\frac{5}{14}$	=	$\frac{7}{14}$
+		+		+
$\frac{3}{14}$	+	$\frac{2}{14}$	=	$\frac{5}{14}$
=		=		=
$\frac{5}{14}$	+	$\frac{7}{14}$	=	$\frac{12}{14}$

Challenge

Fraction Puzzle

The fraction puzzle below has 21 pieces. The pieces are not all the same size. Trace the fraction bars. Be sure to make the number of each size of fraction bar that is shown to the left of the bars. Cut the bars out and label them with the correct fractions. Arrange the pieces on the fraction puzzle so that the 21 fraction bars fit on the puzzle. Glue the pieces to the puzzle. **Possible arrangement is given.**

one: | 1 |

one: | $\frac{1}{2}$ |

two: | $\frac{1}{3}$ |

two: | $\frac{1}{4}$ |

four: | $\frac{1}{6}$ |

five: | $\frac{1}{5}$ |

six:

1				
$\frac{1}{4}$	$\frac{1}{4}$	$\frac{1}{6}$	$\frac{1}{3}$	
$\frac{1}{9}$ $\frac{1}{9}$ $\frac{1}{9}$	$\frac{1}{6}$	$\frac{1}{2}$		
$\frac{1}{5}$	$\frac{1}{5}$	$\frac{1}{5}$	$\frac{1}{5}$	$\frac{1}{5}$
$\frac{1}{3}$	$\frac{1}{6}$	$\frac{1}{6}$	$\frac{1}{9}$ $\frac{1}{9}$	$\frac{1}{9}$

CW104 Challenge

Name _____

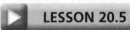

Find the Difference

Fill in the empty squares to have correct subtraction across and down.

1.

$\frac{6}{6}$	−	$\frac{3}{6}$	=	$\frac{3}{6}$
−		−		−
$\frac{4}{6}$	−	$\frac{2}{6}$	=	$\frac{2}{6}$
=		=		=
$\frac{2}{6}$	−	$\frac{1}{6}$	=	$\frac{1}{6}$

2.

$\frac{9}{10}$	−	$\frac{3}{10}$	=	$\frac{6}{10}$
−		−		−
$\frac{5}{10}$	−	$\frac{2}{10}$	=	$\frac{3}{10}$
=		=		=
$\frac{4}{10}$	−	$\frac{1}{10}$	=	$\frac{3}{10}$

3.

$\frac{9}{12}$	−	$\frac{4}{12}$	=	$\frac{5}{12}$
−		−		−
$\frac{6}{12}$	−	$\frac{2}{12}$	=	$\frac{4}{12}$
=		=		=
$\frac{3}{12}$	−	$\frac{2}{12}$	=	$\frac{1}{12}$

4.

$\frac{8}{9}$	−	$\frac{3}{9}$	=	$\frac{5}{9}$
−		−		−
$\frac{4}{9}$	−	$\frac{1}{9}$	=	$\frac{3}{9}$
=		=		=
$\frac{4}{9}$	−	$\frac{2}{9}$	=	$\frac{2}{9}$

5.

$\frac{18}{20}$	−	$\frac{10}{20}$	=	$\frac{8}{20}$
−		−		−
$\frac{9}{20}$	−	$\frac{2}{20}$	=	$\frac{7}{20}$
=		=		=
$\frac{9}{20}$	−	$\frac{8}{20}$	=	$\frac{1}{20}$

6.

$\frac{12}{14}$	−	$\frac{5}{14}$	=	$\frac{7}{14}$
−		−		−
$\frac{4}{14}$	−	$\frac{1}{14}$	=	$\frac{3}{14}$
=		=		=
$\frac{8}{14}$	−	$\frac{4}{14}$	=	$\frac{4}{14}$

Challenge CW105

Name _____

LESSON 20.6

Solve the Riddle

Look at the bottom of the page for the answer to each subtraction problem. Write on the line above the answer the circled letter that is next to the subtraction problem. (Some letters appear more than once.)

1. $\dfrac{8}{10} - \dfrac{3}{10} = \underline{\dfrac{5}{10}}$ Ⓞ

2. $\dfrac{11}{12} - \dfrac{1}{12} = \underline{\dfrac{10}{12}}$ Ⓢ

3. $\dfrac{9}{12} - \dfrac{1}{12} = \underline{\dfrac{8}{12}}$ Ⓔ

4. $\dfrac{5}{8} - \dfrac{4}{8} = \underline{\dfrac{1}{8}}$ Ⓤ

5. $\dfrac{5}{6} - \dfrac{3}{6} = \underline{\dfrac{2}{6}}$ Ⓣ

6. $\dfrac{8}{12} - \dfrac{6}{12} = \underline{\dfrac{2}{12}}$ Ⓝ

7. $\dfrac{7}{8} - \dfrac{5}{8} = \underline{\dfrac{2}{8}}$ Ⓡ

8. $\dfrac{4}{5} - \dfrac{2}{5} = \underline{\dfrac{2}{5}}$ Ⓜ

9. $\dfrac{6}{10} - \dfrac{4}{10} = \underline{\dfrac{2}{10}}$ Ⓟ

10. $\dfrac{8}{8} - \dfrac{2}{8} = \underline{\dfrac{6}{8}}$ Ⓐ

N	U	M	E	R	A	T	O	R	S
$\dfrac{2}{12}$	$\dfrac{1}{8}$	$\dfrac{2}{5}$	$\dfrac{8}{12}$	$\dfrac{2}{8}$	$\dfrac{6}{8}$	$\dfrac{2}{6}$	$\dfrac{5}{10}$	$\dfrac{2}{8}$	$\dfrac{10}{12}$

A	R	E
$\dfrac{6}{8}$	$\dfrac{2}{8}$	$\dfrac{8}{12}$

T	O	P	S
$\dfrac{2}{6}$	$\dfrac{5}{10}$	$\dfrac{2}{10}$	$\dfrac{10}{12}$

CW106 Challenge

Musical Math

These problems have all been solved by the Musical Mathematicians. Their answers are not all reasonable. Read the problems. Decide if the answers given are reasonable. If an answer is not reasonable, explain why.
Possible explanations are given.

1. During a break, some of the band members shared a pizza. The singer and the drummer each ate $\frac{3}{12}$ of the pizza. The rest of the pizza was saved until the next break. How much of the pizza was saved?

   $\frac{3}{12}$ was saved. Since $\frac{3}{12} + \frac{3}{12} = \frac{6}{12}$, then $\frac{6}{12}$ of the pizza was eaten, so $\frac{6}{12}$ is still there. The answer is not reasonable.

2. A band played for a party. $\frac{4}{10}$ of the songs they played were Country Western, $\frac{3}{10}$ were Rock and Roll, and the rest were Jazz. What fraction of their songs were Jazz?

   $\frac{3}{10}$ were Jazz.

   The answer is reasonable.

3. The guests at the party were asked if they liked the band's music. $\frac{5}{10}$ said they did and $\frac{4}{10}$ of the guests said they did not. The other guests could not decide. What part of the guests at the party could not decide?

   $\frac{2}{5}$ could not decide.

   Since $\frac{5}{10} + \frac{4}{10} = \frac{9}{10}$ and $\frac{10}{10} - \frac{9}{10} = \frac{1}{10}$, the answer is not reasonable.

4. The room where the party was held was divided into sections. The band used $\frac{1}{6}$ of the room. The food tables used $\frac{1}{6}$ of the room. The rest of the room was used by the guests. How much of the room was used by the guests?

   $\frac{4}{6}$ was used by the guests.

   The answer is reasonable.

Name _____

LESSON 21.1

Riddlegram!

Answer this riddle. Write the letter that matches each fraction or decimal. You will use some models more than once.

Riddle: Why do you measure snakes in inches?

B E C A U S E $\dfrac{T}{0.2}$ $\dfrac{H}{\frac{7}{10}}$ $\dfrac{E}{\frac{5}{10}}$ $\dfrac{Y}{0.9}$

$\dfrac{H}{0.7}$ $\dfrac{A}{\frac{1}{10}}$ $\dfrac{V}{0.4}$ $\dfrac{E}{\frac{5}{10}}$ $\dfrac{N}{0.3}$ $\dfrac{O}{\frac{8}{10}}$ $\dfrac{F}{\frac{6}{10}}$ $\dfrac{E}{0.5}$ $\dfrac{E}{0.5}$ $\dfrac{T}{\frac{2}{10}}$!

T
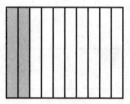

E
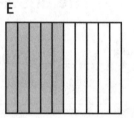

Y
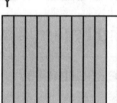

N
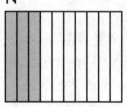

O
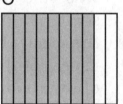

V
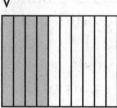

H
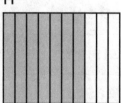

A
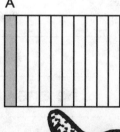

F
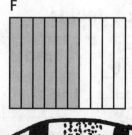

CW108 Challenge

Name _____

LESSON 21.2

Add It Up

Each circle or bar is divided into tenths. Use two crayons in different colors to show two decimal numbers that equal one whole. Then complete the number sentence below the picture to show how you have made one whole. The first one has been done for you.

Check students' drawings and number sentences.

1.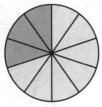
2.
3.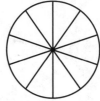

   0.3 + 0.7 = 1.0   ___ + ___ = 1.0   ___ + ___ = 1.0

4.   ___ + ___ = 1.0
5.   ___ + ___ = 1.0

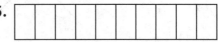

For each circle or bar below, use three crayons in different colors to show three decimal numbers that equal one whole. Then complete the number sentence to show how you have made one whole.

Check students' drawings and number sentences.

6.
7.

   ___ + ___ + ___ = 1.0   ___ + ___ + ___ = 1.0

For each circle or bar below, use four crayons in different colors to show four decimal numbers that equal one whole. Then complete the number sentence to show how you have made one whole. **Check students' drawings and number sentences.**

8.
9.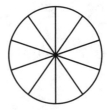

   ___ + ___ + ___ + ___ = 1.0   ___ + ___ + ___ + ___ = 1.0

Challenge   CW109

Sum Match

Find each sum. Draw a line to match.

C	0.3 + 0.4	0.8
Y	0.5 + 0.1	0.2
H	0.1 + 0.1	1.0
O	0.7 + 0.1	0.7
O	0.2 + 0.1	0.6
B	0.6 + 0.4	0.1
E	0.2 + 0.3	0.4
A	0.0 + 0.1	0.3
M	0.7 + 0.2	0.5
N	0.1 + 0.3	0.9

What do bees use to comb their hair?

A H O N E Y   C O M B
0.1 0.2 0.3 0.4 0.5 0.6   0.7 0.8 0.9 1.0

LESSON 21.3

CW110 Challenge

Name _____

LESSON 21.4

Decimal Differences

Find each difference. Draw a line to match.

Y	0.9 − 0.4		0.7
U	0.5 − 0.3		0.3
E	0.8 − 0.1		0.5
M	0.7 − 0.4		0.2
G	0.3 − 0.2		0.1
B	0.9 − 0.3		0.0
M	0.6 − 0.2		0.8
A	0.5 − 0.5		0.9
R	0.9 − 0.0		0.6
A	0.9 − 0.1		0.4

What do you call a bear with no teeth?

A	G	U	M	M	Y	B	E	A	R
0.0	0.1	0.2	0.3	0.4	0.5	0.6	0.7	0.8	0.9

Challenge CW111

Name _____

LESSON 21.5

Digits and Decimals

1. Write all of the decimals you can make by placing the digits 1, 2, and 3 in each of the boxes below. Draw a ring around the greatest decimal you make. Underline the least decimal you make. **The order of answers will vary.**

   <u>1 . 2 3</u>   1 . 3 2

   2 . 1 3   2 . 3 1

   3 . 1 2   (3 . 2 1)

2. Write all of the decimals you can make by placing the digits 2, 4, and 6 in each of the boxes below. Draw a ring around the greatest decimal you make. Underline the least decimal you make. **The order of answers will vary.**

   <u>2 4 . 6</u>   2 6 . 4

   4 2 . 6   4 6 . 2

   6 2 . 4   (6 4 . 2)

For Exercises 3–10, arrange digits in the boxes to make number sentences that are true. Use the digits 1, 3, and 5 in each number sentence.

3. $1.00 < \boxed{1}.\boxed{3}\boxed{5} < 1.50$

4. $3.00 < \boxed{3}.\boxed{1}\boxed{5} < 3.40$

5. $3.10 < \boxed{3}.\boxed{1}\boxed{5} < 3.50$

6. $5.00 < \boxed{5}.\boxed{1}\boxed{3} < 5.20$

7. $10.0 < \boxed{1}\boxed{3}.\boxed{5} < 14.0$

8. $30.0 < \boxed{3}\boxed{1}.\boxed{5} < 35.0$

9. $50.0 < \boxed{5}\boxed{1}.\boxed{3} < 53.0$

10. $1.50 < \boxed{1}.\boxed{5}\boxed{3} < 1.60$

Challenge

Name _____

LESSON 22.1

Multiplication Patterns

You know how to use mental math to multiply any number times 100. Look how you can use mental math to multiply times 99.

Just multiply by 100 and then subtract the original number.
$$8 \times 99 = (8 \times 100) - 8$$
$$= 800 - 8$$
$$= 792$$

Complete.

1. $7 \times 99 = (7 \times \underline{100}) - 7$
   $= 700 - \underline{7}$
   $= 693$

2. $4 \times 99 = (4 \times \underline{100}) - \underline{4}$
   $= 400 - \underline{4}$
   $= 396$

3. $5 \times 99 = (\underline{5} \times \underline{100})$
   $- \underline{5}$
   $= 500 - \underline{5}$
   $= \underline{495}$

4. $6 \times 99 = (\underline{6} \times \underline{100})$
   $- \underline{6}$
   $= \underline{600} - \underline{6}$
   $= \underline{594}$

Use mental math to solve.

5. $3 \times 99 = \underline{297}$
6. $2 \times 99 = \underline{198}$
7. $1 \times 99 = \underline{99}$
8. $9 \times 99 = \underline{891}$
9. $10 \times 99 = \underline{990}$
10. $11 \times 99 = \underline{1,089}$

For 11–19, use the answers to Problems 1–9 to help.
$$80 \times 99 = (80 \times 100) - 80$$
$$= 8,000 - 80$$
$$= 7,920$$

11. $70 \times 99 = \underline{6,930}$
12. $40 \times 99 = \underline{3,960}$
13. $50 \times 99 = \underline{4,950}$
14. $60 \times 99 = \underline{5,940}$
15. $30 \times 99 = \underline{2,970}$
16. $20 \times 99 = \underline{1,980}$
17. $80 \times 99 = \underline{7,920}$
18. $90 \times 99 = \underline{8,910}$
19. $100 \times 99 = \underline{9,900}$

Challenge

Name _____

LESSON 22.2

In the Shade

Estimate the first factor. Then find the estimated product. Draw lines to connect the dots from the smallest product to the largest product.

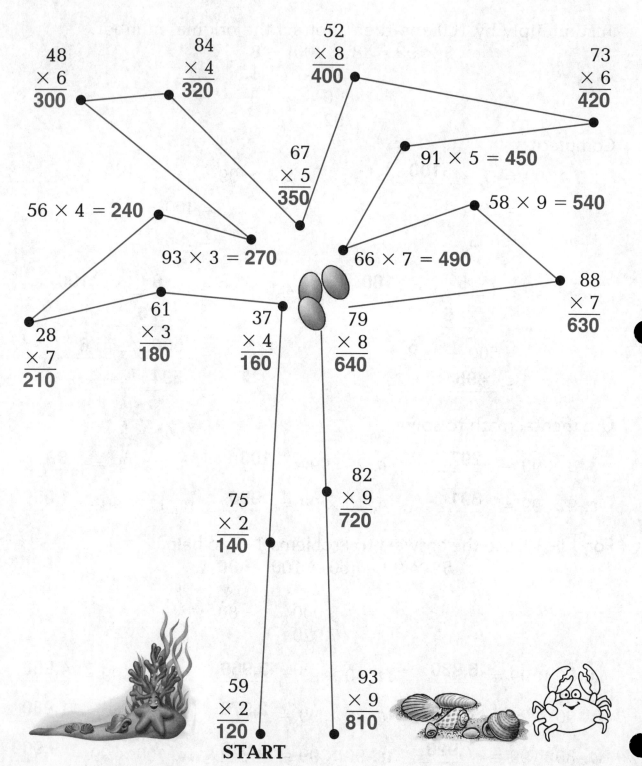

CW114   Challenge

Name _____

LESSON 22.3

Where's the Match?

Use base-ten blocks to find each product. Draw lines to connect the socks with matching products.

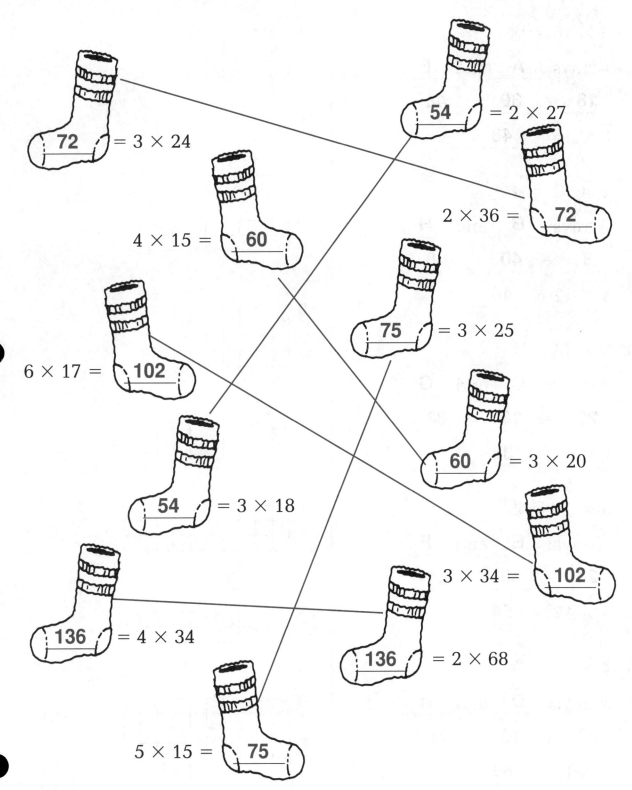

Challenge CW115

Name _____

LESSON 22.4

Pick a Pair

Use the Distributive Property to choose two arrays for each problem. Write the letter for each array. Then use the partial products to solve.

1. 3 × 16 = ■

   Arrays: __A__ and __F__

   __18__ + __30__ = __48__

   3 × 16 = __48__

2. 4 × 12 = ■

   Arrays: __B__ and __H__

   __8__ + __40__ = __48__

   4 × 12 = __48__

3. 2 × 19 = ■

   Arrays: __C__ and __G__

   __20__ + __18__ = __38__

   2 × 19 = __38__

4. 3 × 18 = ■

   Arrays: __E__ and __F__

   __24__ + __30__ = __54__

   3 × 18 = __54__

5. 4 × 17 = ■

   Arrays: __D__ and __H__

   __28__ + __40__ = __68__

   4 × 17 = __68__

A

B

C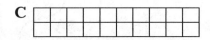

D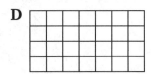

E

F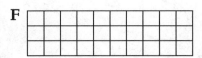

G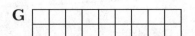

H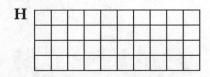

CW116 Challenge

LESSON 22.5

Name _____

Rules and More Rules

Write a multiplication rule for each table. Then complete the table.

1. Rule: __Multiply by 3.__

Factor	Product
14	42
24	72
34	**102**
44	**132**
54	**162**

2. Rule: __Multiply by 5.__

Factor	Product
15	75
20	100
25	**125**
30	**150**
35	**175**

3. Rule: __Multiply by 2.__

Factor	Product
16	32
18	36
20	**40**
22	**44**
24	**48**

4. Rule: __Multiply by 6.__

Factor	Product
11	66
22	132
33	**198**
44	**264**
55	**330**

5. Rule: __Multiply by 4.__

Factor	Product
10	40
13	52
16	**64**
19	**76**
22	**88**
25	**100**
28	**112**

6. Rule: __Multiply by 2.__

Factor	Product
12	24
24	48
36	**72**
48	**96**
60	**120**
72	**144**
84	**168**

Challenge

LESSON 22.6

Name _____

Planning a Picnic

Mrs. Ellis is planning a picnic for 16 people. Use the chart to help her plan the picnic.

Hot dogs	8 in a package
Hot dog buns	4 in a package
Juice	64 ounces in a bottle
Chips	9 servings in a bag
Apples	10 in a bag

1. How many packages of hot dogs would give each person 2 hot dogs?

   4 packages

2. How many packages of hot dog buns would give each person 2 hot dog buns?

   8 packages

3. One bottle of juice will give how many people an 8-ounce serving?

   8 people

4. How many bags of apples would give each person at least one apple?

   2 bags

5. How many bags of chips would give each person at least one serving?

   2 bags

6. How many cookies should Mrs. Ellis bake so that each person can have 3 cookies?

   48 cookies

7. Mrs. Ellis buys a package of 100 napkins. If each person uses 4 napkins, how many napkins will be left after the picnic?

   36 napkins

8. Each picnic table can seat 4 people. How many tables will they need?

   4 tables

CW118 Challenge

Picture This!

Find each product. Match each digit of the product with the picture in the key. Then find the letter that matches the pictures for the product. Write the letter next to the product. Then find the message.

Key

🌼=0   ♥=1   □=2   ○=3   △=4

🌳=5   ⛵=6   ☀=7   📖=8   ◇=9

Example:

```
  127
×   4
-----
  508
```
508 =

1.
```
  265
×   4
-----
  800
```
G

2.
```
  136
×   6
-----
  816
```
R

3.
```
  347
×   4
-----
1,388
```
E

4.
```
  352
×   2
-----
  704
```
A

5.
```
  232
×   5
-----
1,160
```
T

6.
```
  548
×   3
-----
1,644
```
J

7.
```
  421
×   6
-----
2,526
```
O

8.
```
  128
×   3
-----
  384
```
B

9. What is the message?

<u>G</u> <u>R</u> <u>E</u> <u>A</u> <u>T</u>

<u>J</u> <u>O</u> <u>B</u> !

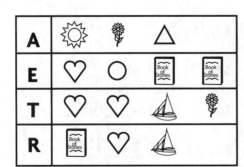

LESSON 22.7

Challenge CW119

Name _____

LESSON 22.8

Can You Find Me?

Use mental math to solve the problems below. Then find the answers in the number search. Circle the answers. You will find them upward, downward, diagonally, and backward.

2	2	0	1	4	7	0	5	1	1	0
5	9	5	3	4	6	4	3	7	2	9
7	4	0	0	2	5	8	3	5	3	8
4	3	9	2	0	0	4	1	3	5	6
6	5	1	4	1	5	5	4	1	8	8
2	1	6	4	0	8	1	7	3	1	5
1	1	1	1	4	2	6	9	5	8	0
4	7	4	5	0	4	2	1	1	1	4
1	8	5	3	3	6	8	5	6	2	3
2	0	1	5	8	5	0	8	0	5	1

1. $56 \times 6 =$ **336**
2. $41 \times 3 =$ **123**
3. $24 \times 3 =$ **72**
4. $58 \times 5 =$ **290**
5. $98 \times 3 =$ **294**
6. $12 \times 8 =$ **96**
7. $54 \times 4 =$ **216**
8. $67 \times 2 =$ **134**
9. $10 \times 7 =$ **70**
10. $26 \times 6 =$ **156**
11. $23 \times 8 =$ **184**
12. $94 \times 2 =$ **188**
13. $71 \times 2 =$ **142**
14. $39 \times 2 =$ **78**
15. $64 \times 5 =$ **320**
16. $15 \times 4 =$ **60**
17. $29 \times 9 =$ **261**
18. $31 \times 5 =$ **155**

CW120 Challenge

Name _____

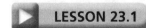 LESSON 23.1

Sharing Marbles

Cassie, Tony, and Mitchell want to share 26 marbles evenly. They decide to put any extra marbles in a jar.

1. How many marbles will each friend get? __8 marbles__

2. How many marbles will they put in the jar? __2 marbles__

Finish this table to show other ways of sharing.

	Number of Marbles	Number of Friends	Marbles for Each Friend	Leftover Marbles
3.	34	4	8	2
4.	29	6	4	5
5.	26	3	8	2
6.	33	6	5	3
7.	14	2	7	0
8.	23	7	3	2
9.	14	4	3	2
10.	22	3	7	1
11.	34	6	5	4

12. A group of 5 friends wants to share a set of stickers evenly. What is the greatest number of stickers that could be left over?

    __4 stickers__

13. A group of friends shares a batch of cookies evenly. There are 3 cookies left over. What is the least number of friends that could be in the group?

    __4 friends__

Challenge CW121

Star Quotients

Find the quotient for each problem. If the quotient matches the number in the star, write the letter on the matching blank below to solve the riddle. You may use base-ten blocks to help.

| 40 ÷ 5 = __8__ | 39 ÷ 3 = __13__ |
| R | E |

| 36 ÷ 9 = __4__ | 28 ÷ 2 = __14__ |
| L | S |

| 81 ÷ 3 = __27__ | 48 ÷ 4 = __12__ |
| E | B |

★ 15

| 64 ÷ 8 = __8__ | 45 ÷ 3 = __15__ |
| W | O |

| 80 ÷ 2 = __40__ | 93 ÷ 3 = __31__ |
| P | T |

| 36 ÷ 2 = __18__ | 72 ÷ 2 = __36__ |
| V | M |

| 42 ÷ 7 = __6__ | 76 ÷ 4 = __19__ |
| D | N |

| 68 ÷ 2 = __34__ | 50 ÷ 5 = __10__ |
| K | E |

Riddle: What starts with E, ends with E, but contains only one letter?

A N __E__ N V __E__ L __O__ __P__ __E__
    10   19   18   27   4   15   40   13

Name _____   **LESSON 23.3**

Arranging Digits

Sumi has written 6 different division problems using only the digits 2, 4, and 6. She is wondering which problems will have the greatest and least quotients. Find the quotient for each problem.

```
            2 3                    (3 2)                    6 r2
1.  2 ) 4 6          2.   2 ) 6 4           3.   4 ) 2 6
      - 4 ↓                  - 6 ↓                  - 2 4
        0 6                    0 4                      2
        - 6                    - 4
          0                      0


            1 5  r2                   4 *                        7
4.  4 ) 6 2          5.   6 ) 2 4           6.   6 ) 4 2
      - 4 ↓                  - 2 4                  - 4 2
        2 2                      0                      0
      - 2 0
          2
```

7. Circle the greatest quotient.

8. Draw a star next to the least quotient.

Write your own division problems by arranging the digits 3, 5, and 7 in different ways. Solve each problem. **Check students' problems and solutions.**

```
              1 9                      2 5                       7 r2
9.  3 ) 5 7          10.  3 ) 7 5           11.  5 ) 3 7


              1 4 r3                     5                       7 r4
12. 5 ) 7 3          13.  7 ) 3 5           14.  7 ) 5 3
```

Challenge CW123

Name _____

LESSON 23.4

Colorful Quotients

Use mental math to divide. Write the answer inside each triangle.

Use a crayon to color the triangles that have an even-numbered quotient blue.

Color the triangles with an odd-numbered quotient yellow.

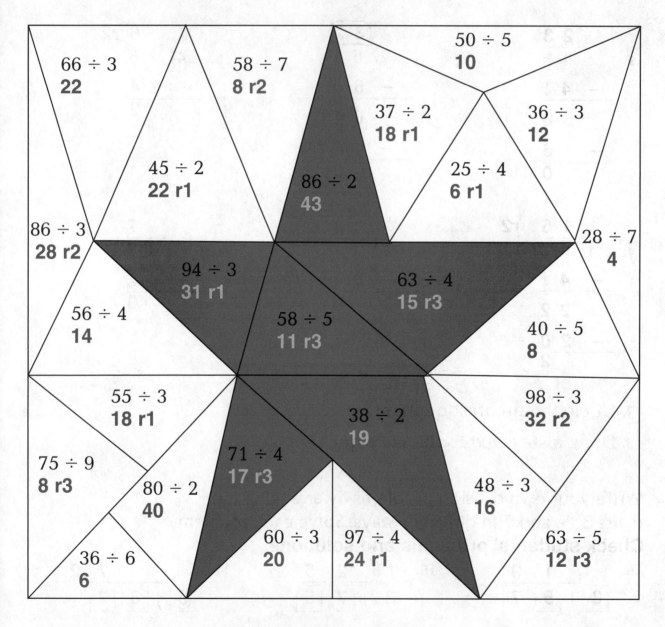

1. What shape is colored yellow? ___a star___
2. How many quotients were even? ___20___
3. How many quotients were odd? ___6___

CW124 Challenge

Name _____

LESSON 23.5

The Division Race

Choose a partner.

Materials: number cube labeled 1–6, game token for each player, paper, and pencil

How to Play: Place game tokens on Start. Take turns tossing the number cube. Use the number tossed as the divisor for the number on your space. Use paper and pencil to divide. Then, move the number of spaces shown in the remainder. If there is no remainder, do not move. The first player to cross the finish line is the winner.

Example: Player's game token is on 86. Player tosses 3. 86 ÷ 3 = 28 r2. Move 2 spaces.

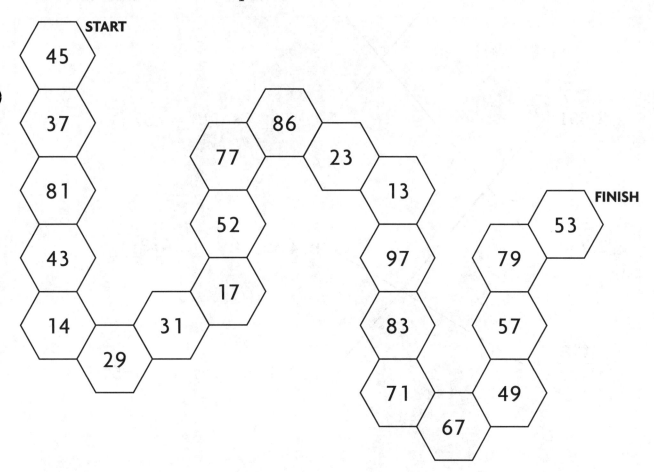

Challenge CW125

Divide and Check

Solve each division problem. Then complete the number sentence that can be used to check the answer. Draw a line from the division problem to the related number sentence.

1. 4)452̄ quotient **113**

2. 5)805̄ quotient **161**

3. 3)651̄ quotient **217**

4. 2)758̄ quotient **379**

5. 4)472̄ quotient **118**

A. __5__ × 161 = __805__

B. __2__ × 379 = __758__

C. __4__ × 113 = 452

D. 4 × 118 = __472__

E. __3__ × 217 = __651__

CW126 Challenge

Name _____

LESSON 23.7

Something's Fishy!

Divide each number by 2 and by 3. Use mental math, pictures, or models. Then write each number where it belongs in the Venn diagram below.

Numbers Evenly Divided by 2   **Numbers Evenly Divided by 3**

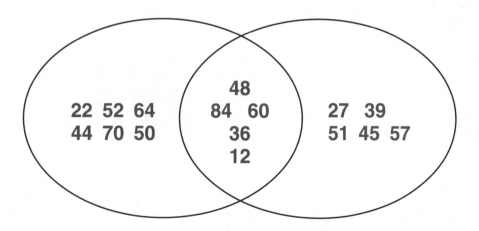

22 52 64
44 70 50

48
84 60
36
12

27 39
51 45 57

The numbers in the overlapping part of the Venn diagram can be divided evenly by two other numbers that are less than ten. What are the numbers? __4, 6__

Challenge CW127